STOCHASTIC HYDROLOGY

STOCHASTIC HYDROLOGY

By

DR. P. JAYARAMI REDDY

Principal

G. Pulla Reddy Engineering College

Kurnool-518002

LAXMI PUBLICATIONS (P) LTD

(An ISO 9001:2015 Company)

BENGALURU • CHENNAI • GUWAHATI • HYDERABAD • JALANDHAR

KOCHI • KOLKATA • LUCKNOW • MUMBAI • RANCHI

NEW DELHI

Printed and bound in India

First Published: 1987, Second Editio : 2016

New Edition : 2025

ISBN : 978-81-318-0983-9

Branches

✆	Bengaluru	080-26 75 69 30
✆	Chennai	044-24 34 47 26
✆	Guwahati	0361-254 36 69
✆	Hyderabad	040-27 55 53 83
✆	Kochi	0484-405 13 03
✆	Kolkata	033-40 04 77 79
✆	Lucknow	0522-430 36 13
✆	Ranchi	0651-224 24 64

PUBLISHED IN INDIA BY

LAXMI PUBLICATIONS (P) LTD

An ISO 9001:2015 Company

113, GOLDEN HOUSE, GURUDWARA ROAD, DARYAGANJ,

NEW DELHI - 110002, INDIA

Corporate Office : B-33, Sector-5, Noida,

Gautambuddha Nagar (U.P.) - 201 301, INDIA

Telephone : +91-120-44 46 900

www.laxmipublications.com

info@laxmipublications.com

To the Memory of my Father
BUGGA RAMI REDDY

PREFACE TO THE SECOND EDITION

In bringing out the second edition of the book, the opportunity has been utilised to thoroughly revise and enlarge the subject matter taking into account the suggestions offered by many. Few mistakes which crept into the first edition have been rectified. Solved examples have been increased manifold. And in this edition all chapters are provided with Exercises at the end to help the readers to test their comprehension of the subject.

Suggestions for further improvement of the book will be thankfully received.

P. JAYA RAMI REDDY

Kurnool
January, 1997

PREFACE TO THE FIRST EDITION

The importance of stochastic modelling of hydrological variables like runoff and precipitation in the design of water resources systems is too well recognised to need any further emphasis. Most of the Indian Universities offering Postgraduate programmes in Hydraulics and Water Resources Engineering have included "Stochastic Hydrology" as one of the subjects in their syllabi. The material presented in this text-book is the outcome of teaching the subject to postgraduate students at Regional Engineering College, Warangal, for over five years and the extensive study and use of stochastic models in my doctoral work.

After a brief introduction in Chapter 1, the fundamentals of probability theory are developed in Chapter 2. Most of the important probability distributions of both discrete and continuous variables, which are frequently used to model the hydrologic phenomena are described in Chapter 3. Chapter 4 concentrates on methods of estimation and fitting appropriate distributions to the observed data, while Chapter 5 covers the topics of correlation and regression. All these chapters form the basic background to the understanding and the appreciation of the principles of stochastic processes and their applications in hydrology that are discussed in the subsequent chapters. Chapter 6 deals with the definitions and classifications of stochastic processes and time series. Chapters 7 and 8 are, respectively, devoted to the two important analysis techniques, namely, the autocorrelation analysis and the spectral analysis. Synthetic flow generation models are elaborately discussed in Chapter 9. Finally Chapter 10 contains a brief description of the methods of generation of random numbers.

Readers already familiar with probability theory can perhaps, skip the Chapters 2 through 5. But the material presented in this text-book is application oriented and is quite different from the presentations found in the other text-books on probability written by mathematicians. Most of the numerical examples are solved on a pocket calculator. The data for the correlogram and spectral density of the examples in Chapters 7 and 8 are obtained through computer analysis. When the generation of synthetic flows is required for longer periods, one has to necessarily resort to the computer. As the book stresses more on applications, several results, especially in autocorrelation analysis and spectral analysis have been taken for granted, and more or less throughout the book, a compromise has been struck between mathematical rigour and methodology.

I express my gratitude to the authors of the various books and technical papers listed in the bibliography, which I consulted. I am indebted to all my colleagues and postgraduate students of the Division of Water Resources at the Regional Engineering College, Warangal for their encouragement and help in writing this book, In particular, I would like to thank professor R. Raghavendran who has been a source of inspiration for me.

I express my gratefulness to Sri R.K. Gupta for his efforts in bringing out this book.

Suggestions for improvement shall be gratefully acknowledged.

P. JAYARAMI REDDY

Vidyanagar
September, 1986.

CONTENTS

			Pages
1.	**INTRODUCTION**		**1—4**
	Exercises		3
2.	**PROBABILITY**		**5—40**
2.1.	Introduction		5
2.2.	Set Theory	...	6
	Events and Sets	...	7
	Algebra of sets	...	9
2.3.	Axioms of Probability	...	14
2.4.	Interpretation of Probability		15
	The classical interpretation		15
	The frequency interpretation		15
	The subjective interpretation		15
2.5.	Elementary Properties of Probability		16
2.6.	Conditional Probability		18
2.7.	Total Probability and Bayes' Theorem		22
2.8.	Random Variable		23
2.9.	Probability Function and Distribution Function of a Discrete Random Variable		24
2.10.	Probability Density Function and Distribution Function of a Continuous Random Variable		26
2.11.	Distribution Function of a Mixed Random Variable		28
2.12.	Bivariate and Multivariate Distributions	...	30
2.13.	Marginal Distributions	...	32
2.14.	Conditional Distributions	...	33
2.15.	Derived Distributions	...	35
	Exercises	...	37
3.	**SOME CHARACTERISTICS OF DISTRIBUTIONS**		**41—54**
3.1.	Introduction		41
3.2.	Mean		41
3.3.	Variance		43
3.4.	Chebyshev's Inequality		44
3.5.	Covariance		46

			Pages
	3.6.	Moments	48
		Skewness coefficient	49
		Kurtosis coefficient	50
	3.7.	Moment Generating Function	52
		Exercises	53
4.	**CONTINUOUS DISTRIBUTIONS**		**55—73**
	4.1.	Introduction	55
	4.2.	Normal Distribution	55
		Approximations to F(z)	58
		Central limit theorem	59
	4.3.	Log-Normal Distribution	61
	4.4.	Gamma Distribution	64
		Special Cases of Gamma Distribution	66
	4.5.	Beta Distribution	67
	4.6.	Uniform Distribution	67
	4.7.	Extreme Value Distribution	68
		Extreme Value Type I Distribution for Largest Value (Gumbel)	68
		Extreme Value Type III Distribution for Smallest Value (Weibull)	70
	4.8.	Pearson Distributions	71
		Pearson Type III Distribution	71
		Log-Pearson Type III Distribution	71
	4.9.	Other Continuous Distributions	72
		Exercises	72
5.	**DISCRETE DISTRIBUTIONS**		**74—85**
	5.1.	Binomial Distribution	74
	5.2.	Negative Binomial Distribution	79
	5.3.	Geometric Distribution	80
	5.4.	Poisson Distribution	81
	5.5.	Hypergeometric Distribution	83
	5.6.	Other Discrete Distributions	84
		Exercises	84
6.	**ESTIMATION AND PROBABILITY PLOTTING**		**86—114**
	6.1.	Sample Statistics	86
	6.2.	Estimation of the parameters of the Distribution	88
		Method of Moments	89
		Method of Maximum Likelihood	89
	6.3.	Selection of Distribution	91

			Pages
	6.4.	Probability Plotting	93
		California formula	94
		Hazen's formula	94
		Weibull's formula	94
	6.5.	Chi-Squared Test	103
	6.6.	Smirnov-Kolmogorov Test	107
	6.7.	Frequency factors	108
		Normal Distribution	109
		Gumbel's Distribution	110
		Log-Pearson Type III Distribution	110
		Exercises	112
7.	**CORRELATION AND REGRESSION**		**115—133**
	7.1.	Correlation	115
		Spurious Correlation	118
		Coefficient of Determination	118
		Rank Correlation Coefficient	120
	7.2.	Regression	122
		Simple Linear Regression	123
		Transformations	125
		Polynomial fitting	126
	7.3.	Multiple Regression	127
		Measures of Adequacy	130
		Selection of Variables	130
		Exercises	132
8.	**STOCHASTIC PROCESSES**		**134—149**
	8.1.	Introduction	134
	8.2.	Classification of Stochastic Processes	134
	8.3.	Stationary Process	137
		Tests of Stationarity	140
	8.4.	Description of a Stochastic Process	141
	8.5.	Classification of Time Series	141
		Sinusoidal series	142
		Complex periodic series	142
		Almost periodic series	144
		Transient series	145
		Stationary series	146
		Non-stationary series	146
	8.6.	Components of time series	147
		Methods of Investigation	147
		Exercises	149

		Pages
9.	**AUTOCORRELATION ANALYSIS**	**150—171**
9.1.	Estimation of the Autocorrelation Coefficient	150
9.2.	Correlogram of an Independent Process	152
9.3.	Moving Average Process	155
9.4.	Auto Regressive Process	158
	First order AR Process	160
	Second order AR Process	162
9.5.	Autoregressive Moving Average Process	164
	ARMA (1, 1) Process	165
9.6.	Autoregressive Integrated Moving Average Process	166
9.7.	Correlograms of other Processes	166
	Exercises	170
10.	**SPECTRAL ANALYSIS**	**172—189**
10.1.	Introduction	172
10.2.	Line spectrum	172
10.3.	Continuous spectrum	176
10.4.	Smoothing of Spectral Density Estimates	178
	Bartlett's Window	178
	Parzen's Window	178
	Blackman and Tukey (Hanning) Window	179
10.5.	Spectral Density function of an Independent Process	180
10.6.	Spectrum of a Moving Average Process	185
10.7.	Spectrum of an Autoregressive Process	185
	AR (1) Process	185
	AR (2) Process	186
10.8.	Spectrum of Autoregressive Moving Average Process	187
10.9.	Spectrums of other Processes	187
	Exercises	189
11.	**UNIVARIATE STOCHASTIC MODELS**	**190—217**
11.1.	Introduction	190
11.2.	Classification of Models	191
11.3.	Univariate Annual Models with Normal Distribution	192
	11.3.1. First order AR Model	193

			Pages
	11.3.2.	Second order AR Model	195
	11.3.3.	ARMA Model	197
11.4.	Univariate Annual Models with Non-Normal Distribution		199
	Gamma Distribution		200
	Lognormal Distribution		201
11.5.	Univariate Annual Models obeying Hurst's Law		202
	11.5.1.	Fractional Gaussian Noise Models	205
		Type II Approximation to fGn	205
		Filtered fGn Models	205
		Fast Fractional Gaussian Noise Models	206
11.6.	Univariate Seasonal Models		207
	11.6.1.	Thomas-Fiering Model	207
		Gamma Distribution	209
		Lognormal Distribution	210
	11.6.2.	Daily Flow Model	211
		Exercises	216
12.	**MULTIVARIATE STOCHASTIC MODELS**		**218—230**
12.1.	Introduction		218
12.2.	Multisite Annual Models		219
	Multisite First order AR Model		219
	Multisite Higher order AR Model		227
12.3.	Multisite AR Model for Seasonal Flows		228
	Non-stationary Multivariate Seasonal Model		229
12.4.	Other Types of Multisite Models		229
	Exercises		229
13.	**GENERATION OF RANDOM VARIATES**		**231—245**
13.1.	Introduction		231
13.2.	Uniformly Distributed Random Numbers		232
	Mid-Square Technique		232
	Mid-Product Technique		233
	GPSS Algorithm		233
	The multiplicative Congruential Method		233
	Mixed congruential Method		234
	Testing the random number sequence		235
13.3.	Generation of Normal Random Numbers		237

			Pages
	The Inverse Transformation Method		237
	The Central Limit Theorem Method		238
	Box-Muller Method		239
13.4.	Random Numbers of Other Distributions		243
	Exponential Distribution		244
	Gamma Distribution		244
	Log-Normal Distribution		244
	Binomial Distribution		244
	Poisson Distribution		245
	Exercises		245
	Bibliography	...	**246**
	Appendix	...	**249**
	Index	...	**258**

1

Introduction

Water resources engineering, which is one of the many activities of a civil engineer, aims at transforming the water resources vector as it occurs in nature into a most desirable one with respect to time, space, quantity and quality. In other words, the water resources projects are executed and operated with the purpose of making water available for various uses at the times when it is needed, at places where it is needed, in adequate quantities and with an acceptable quality. This is required because of highly non-uniform distribution of natural water resources on the globe earth with respect to both time and space. The water resources systems are very complex due to the variety of objectives, conflicting nature of water uses, and their impact on socio-political environment. The design of water resources systems is further complicated because of the fact that the hydrological variables like precipitation, runoff, evaporation etc., which are essentially required in the assessment of available water resources of the region as well as demand for water for various purposes like irrigation, power generation, navigation etc., are in general random in nature. Even under seemingly identical conditions the outcomes of these variables are different during different observations, precluding their deterministic prediction. If these variables are pure random they can be modelled by a single probability distribution function which does not vary with time. Unfortunately most of these variables are stochastic quantities with their probability law being time variant making the sequence of their occurrence more important. Therefore, any water resources project whose design is based on averages of hydrological data is likely to be a failure either commercially because the capacities of the components provided in the system such as spillway, reserviors etc., are so large in size than the corresponding capacities which are actually required during the operation in its life period, or structurally because the components may be found wanting in their capacities to handle the quantities of water which are realised during its life period. Also, the operation policy based on soundly predicted inflows into the reservoirs of the system rather than on average flows will enhance the system's utility manifold.

The principles of Stochastic Processes, a branch of Probability and Mathematical Statistics and of recent origin, have been increasingly and successfully applied in the past three decades to model many of the hydrological processes which are realised to be stochastic in nature. The advent of computers made the application of these stochastic techniques possible with ease and speed. Stochastic methods try to extract maximum possible information from the available record and aid the designer in evaluating his designs more objectively. Consider, for example, the fixation of the storage capacity of a single reservoir system. The conventional approach is to employ the mass curve technique using the single sequence of available inflows and arrive at the required storage capacity for any specified demand pattern. There are many obvious drawbacks in this method. The fault lies not with the mass curve technique but with the deterministic sequence of inflows and the deterministic demand pattern used in the analysis. First, the length of the record may be less than the projected useful life period of the reservoir. Secondly, even if the length of the record is equal to the life period of the project, the answer for the storage capacity is correct only when the same sequence of inflows and the same deterministic demand pattern are realised during its operation which is seldom the case. Thirdly, there is no way of evaluating the risk of failure inherent in the reservoir design as we are dealing with random inflows. Whereas the stochastic methods can analyse the observed data and provide a model which can be used to generate as many equally likely sequences of inflows as required and of the same length as the life period of the project. These sequences possess the same statistical properties as the observed sequence. Each sequence can then be used in the mass curve analysis to yield the storage capacity. Thus we see that the required storage capacity is not a single number. It is a random variable in itself whose probability distribution function can be identified from the answers obtained for this variable corresponding to various sequences used in the mass curve analysis. Now, questions such as what is the probability that the reservoir of a given capacity fails to meet the demand in its life period ? and what should be the capacity of the reservoir if it has to meet the demand with a specified level of assurance in its life period ? can be easily answered. If the demand for water is also treated as a stochastic quantity with due consideration given to its dependence on the inflows, the design will become more realistic. If the system consists of more than one reservoir, stochastic methods become all the more important since the correlation among the flows at various sites has to be incorporated in the design, if it has to be realistic, which is almost impossible in conventional methods.

Stochastic methods can also be used profitably in combination with deterministic watershed simulation models to extend the runoff record. A long sequence of precipitation can be generated using the stochastic model and this when used as input to the watershed simulation model yields a corresponding long runoff sequence thereby increasing the reliability of estimates of runoff.

In this book, an attempt has been made to present the analysis and model building techniques that are commonly used in Stochastic Hydrology. The purpose of the book is to provide the fundamentals to the students and the professional engineers interested in the subject and it does not attempt to report the up-to-date developments as the subject is developing at very rapid rate. The enthusiastic reader is advised to turn to the leading journals in the field. The next six chapters are devoted to develop the requisite background in probability theory, distributions of random variables, estimation and probability fitting methods, and the correlation and regression analysis while the remaining part of the book concentrates on stochastic processes, autocorrelation and spectral analysis techniques, synthetic flow generation models, and random number generation methods. The reader familiar with probability theory can skip the chapters 2 through 7 or can go through them in rapid succession. Wherever possible, the concepts have been illustrated through worked out examples. As this book is meant to be a text-book at graduate level, reasonable acquaintance with the fundamentals of calculus and matrix operations and familiarity with computer have been assumed on the part of the reader through out the discussions.

EXERCISES

1.1. Explain the principle of mass curve technique to find the storage capacity of a reservoir required to meet a demand rate equal to the mean flow.

1.2. What is the *sequent peak algorithm* ? How is storage capacity of a reservoir determined using this algorithm ?

1.3. What are the deficiencies in the answer obtained for the storage capacity by the mass curve analysis using a deterministic sequence of inflows ? How can these deficiencies be overcome ?

1.4. The annual flows in Mm^3 of a river at a particular site for the period 1951 to 1990 are as given below :

625, 855, 743, 1294, 897, 1067, 479, 681, 1124, 1317,
1035, 631, 1126, 681, 909, 1949, 1539, 866, 816, 836,
943, 1000, 1203, 1315, 1309, 794, 2312, 1179, 1051, 979,
488, 1059, 739, 982, 827, 606, 674, 686, 813, 1170.

Determine the storage capacity of a reservoir proposed at this site to meet a demand rate equal to the mean annual flow in the river.

1.5. Determine the storage capacity in the above exercise assuming that the flows in the second row (for the years 1961 to 1970) and the flows in the last row (for the years 1981 to 1990) are interchanged. Is this storage capacity same as the one obtained in Exercise 1.4 ? Comment on the discrepancy.

2

Probability

2.1. INTRODUCTION

Strictly speaking in any physical process there should be a definite relation between the *cause* and the *effect*. Once this relation is known, the outcome of the process can be precisely predicted given the information regarding the factors affecting the process. For example, from the relationship between the head causing the flow through a small orifice H and the velocity through the orifice V, given by the equation $V = \sqrt{2gH}$, we can predict the velocity for any given head. Similarly the declination of the Sun, the time of sunrise and sunset on any day can be uniquely predicted in advance. Such processes are known as *deterministic processes*. But in some processes there is an element of uncertainity or unpredictability regarding the outcome. Under seemingly identical conditions the outcome of the process is different during different observations. For example, though the average weather conditions are the same from year to year the observed rainfall at a given place is not the same every year and it cannot be predicted uniquely. The reasons for our inability to predict the outcome of the process could be that we may not know all the causative factors, we may not have enough data about the initial conditions of the process, the causative factors are so large in number and so complicated that calculation of their combined effect is not feasible, or some of the causative factors themselves are unpredictable. In all such cases, the process is treated as a *random process* or a *probabilistic process* or a *random phenomenon* wherein we would not try to find the deterministic relationship which relates the causative factors and the corresponding effects, but instead we treat the process as a chance phenomenon the outcome of which is governed by some *probability law*. Once this probability law is identified, the probability theory can be used to derive the probabilities of the complex events and statistical averages concerning the process.

One of the more logically satisfying theories of probability is that based on the concept of *set theory*. These concepts are introduced in the next section.

2.2. SET THEORY

Whenever we want to use the probability concepts, we must first define an experiment and specify all possible outcomes of that experiment. The outcomes of the experiments may be often defined numerically. For example, the annual peak discharge of a stream at a particular site might be 756.5 m^3/s. At a specified time of observation the temperature and pressure of the atmosphere at a given observatory might be 28°C and 980.7 millibars. In these situations the outcome of the given experiment can obviously be described geometrically by a point in one or two dimensional space as shown in Fig. 2.1 ; the annual peak discharge can be represented by a point on a line (one dimensional space), whereas the observatory temperature and pressure readings could well be represented by a point on a two dimensional space.

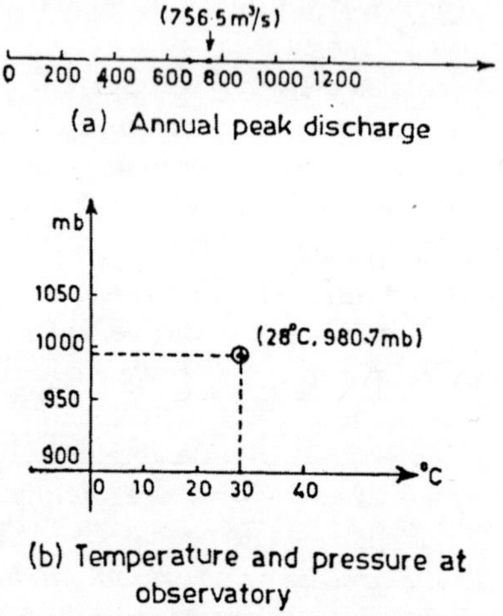

Fig. 2.1. Geometrical representation of experimental outcomes.

In general, if n numbers are required to specify the outcome of a given experiment, then that output can be represented by a point in an n-dimensional space. These points are called the *sample points*. The sample points may also be called the *elements*.

There are some experiments for which the numerical description of the outcome may not be possible or may not be appropriate.

Consider, for example, the flip of a coin. The possible outcomes are *head* and *tail*. Similarly when we pick up a ball from a box containing colour balls, the possible outcomes could be *Red*, *Black*, *White*, *Yellow* etc. Although in these cases the outcome can still be described numerically (for example, representing head by 1 and tail by 0 in the case of flipping a coin, and assigning a value of 1 for Red, 2 for Black, 3 for White and 4 for Yellow etc., in the other experiment of picking up a colour ball), there is no specific way of doing so. However, the idea of geometrically representing the outcome of an experiment as a point in some space can still be used. Thus, as shown in Fig. 2.2 (*a*), a space describing the possible outcomes of flipping a coin can consist of two points ; one labelled H and other labelled T. Similarly, as shown in Fig. 2.2 (*b*), the space describing the possible outcomes of picking a colour ball from a box can consist of points labelled R, B, W, Y etc.

(*a*) Sample points of tossing a coin.

(*b*) Sample points of picking a coloured ball.

Fig. 2.2. Geometrical representation of non-numerical outcomes.

The points which represent the outcomes of an experiment are called the *sample points*. The total space which contains the sample points representing all possible outcomes of an experiment is called the *sample space* of that experiment.

It may be noted here, however, that there can be more than one way of representing the sample point and sample space for a given physical experiment. For example, consider the annual peak discharge of a stream. If we are interested in the actual value of the annual peak discharge, the sample space would be a line representing all real numbers above zero. On the other hand, if what we need to know is whether the annual peak discharge is more than a stipulated value, say 1000 m^3/s, an appropriate sample space would be that consisting of only two sample points, one labelled *above 1000 m^3/s* and the other labelled *below 1000 m^3/s* or more briefly one point labelled *yes* and the other labelled *no*.

Events and Sets. An outcome or a collection of outcomes of a defining experiment is called an *event*. As discussed earlier, an event is represented by either a single sample point or by a collection of sample points in a sample point-sample space representation. A collection of sample points or a grouping of sample points is called a *set*. Thus an event may be represented by a set of sample points in a sample space. The representation of events by sets of sample points will be very useful. The set consisting of all the elements of the sample space is called an *identity set* or the *reference set* and is usually denoted by S.

Let us consider the experiment of observing the number of rainy days in any year at a given place. The sample space for this experiment consists of all integer numbers between 0 and 365 and hence we write

$$S = \{0, 1, 2, \ldots\ldots\ldots, 364, 365\} \quad \ldots(2.1)$$

If A is the event that there are no rainy days in a year, then

$$A = \{0\} \quad \ldots(2.2)$$

That is, A is the set which consists of the single point 0. Similarly, if

$$B = \{101, 102, \ldots\ldots, 364, 365\} \quad \ldots(2.3)$$

then B is the event that the number of rainy days in a year is more than 100. The sample space S and events A and B for this experiment are shown in Fig. 2.3.

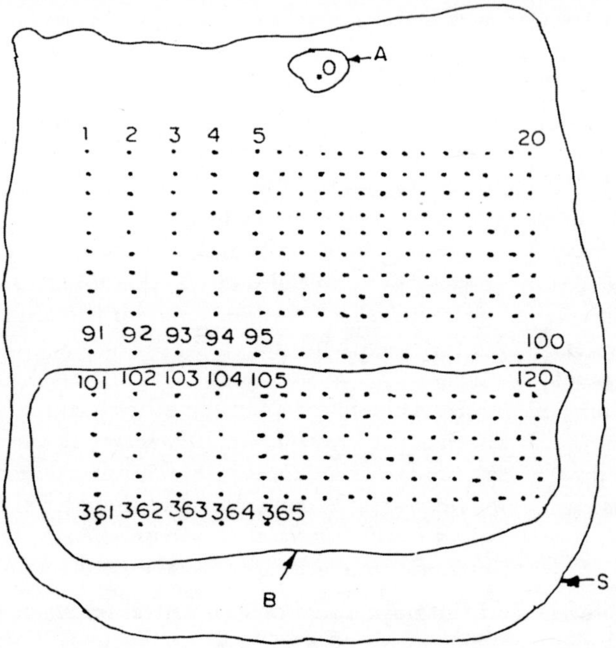

Fig. 2.3. Events A and B and sample space S of the rainy days.

As a second example, consider the experiment of observing the annual rainfall q at a particular raingauge station which is believed not to exceed 1500 mm. A suitable sample space would then be

$$S = \{q : 0 \le q \le 1500\} \qquad ...(2.4)$$

that is, the set of sample points q such that q is greater than or equal to zero and less than or equal to 1500. In this case if event A is defined as

$$A = \{q : 0 \le q \le 750\} \qquad ...(2.5)$$

then A is the event that the annual rainfall is less than or equal to 750 mm. The sample space and event A for this experiment are shown in Fig. 2.4.

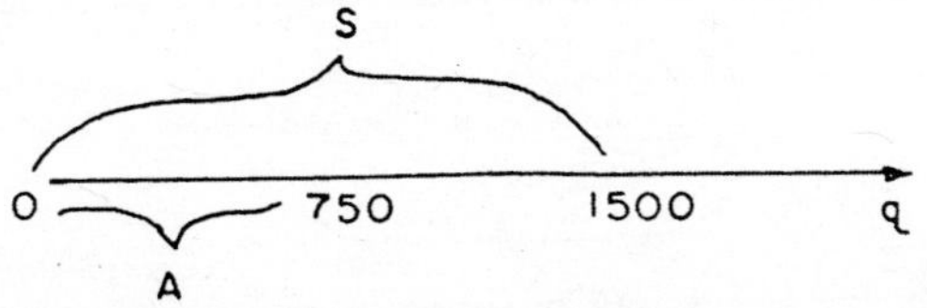

Fig. 2.4. Event A and sample space S for annual rainfall.

An examination of these two examples shows that an event (or set) may be defined either by *listing*, that is by enumeration of its elements as in Eq. (2.1) through (2.3), or by *description*, that is by specification of the properties of its elements as in Eq. (2.4) and (2.5). It should be noted that definition by listing is possible only if the number of sample points in the event is finite or countably infinite. Definition by description is possible in any case, but required if there is an uncountable number of sample points in the event (for, in that case there are too many elements to list-either directly or by induction).

Algebra of sets. Two sets A and B are said to be *identical or equal* if and only if A and B have the same sample points. In such a case we write $A = B$. In other words A and B are two different names for the same set of elements.

If each element of a set A is also an element of a set B, then A is said to be *included* in B and A is called a *subset* of B. This is indicated by writing $A \subset B$. It is often convenient to represent the meaning of relation between sets schematically through a figure. Such a pictorial representation of sets in a sample space is called *Venn Diagram*. Fig. 2.5 is the Venn diagram representing the inclusion $A \subset B$.

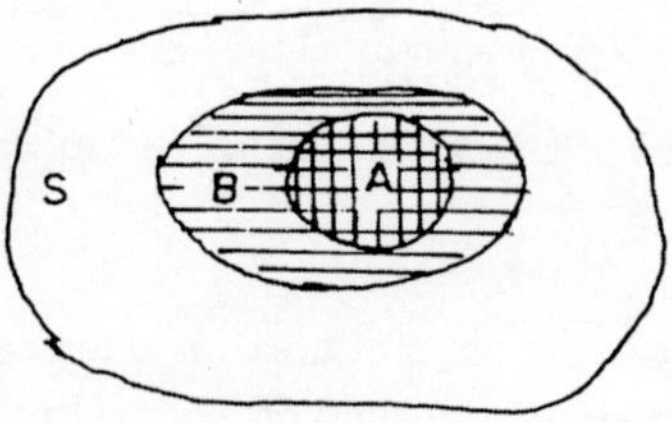

Fig. 2.5. Venn diagram representing the inclusion $A \subset B$.

If a set A has no elements contained in it, then A is callea the *null set* or *vacant set*. This is indicated by writing $A = 0$.

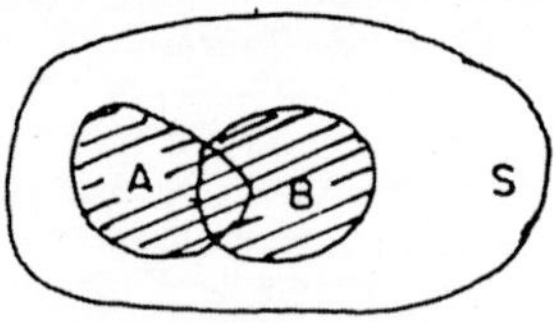

Fig. 2.6. Venn diagram representing the union $A \cup B$.

The set of all elements which belong to atleast one of the sets A and B is called the *union* of A and B. The union of A and B is indicated by writing $A \cup B$. In other words, the union of A and B consists of elements either in set A or in set B or in both. The shaded portion in the Venn diagram of Fig. 2.6 represents the union of A and B. It should be noted here that the sample points which are common to both A and B are to be counted only once. Evidently if A is a subset of B, then the union of A and B is B itself. That is if $A \subset B$, then $A \cup B = B$. Also for every set A, $A \cup A = A$. The union of several sets $A_1, A_2, A_3, ...$ is the set of all elements that belong to at least one of the several sets. This union is denoted by $A_1 \cup A_2 \cup A_3 \cup ......$ If a finite number k of sets is involved $A_1 \cup A_2 \cup ...... \cup A_k$ may also be represented as $\bigcup_{i=1}^{k} A_i$.

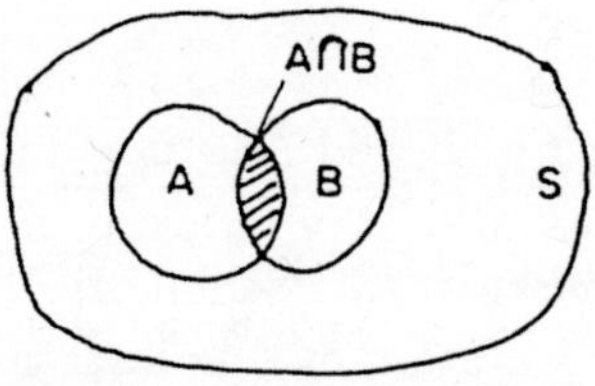

Fig. 2.7. Venn diagram representing the intersection $A \cap B$.

If two sets A and B have some elements which are common to both, then the set made up of those common elements is called the *intersection* of A and B and is denoted by $A \cap B$. The shaded portion in Fig. 2.7 represents the intersection of A and B. Evidently if A is subset of B, then the intersection of A and B is A itself. That is, if $A \subset B$, then $A \cap B = A$. Also for every set A, $A \cap A = A$. The intersection of several sets $A_1, A_2, A_3, \ldots\ldots$ is the set of all elements that belong to each of these sets. This intersection is denoted by $A_1 \cap A_2 \cap A_3 \ldots\ldots$ If a finite number k of sets is involved $A_1 \cap A_2 \cap \ldots\ldots \cap A_k$ may also be represented as $\bigcap_{i=1}^{k} A_i$.

From the definition of union and intersection it follows that for any given two sets A and B the following results hold true.

$$A \subset (A \cup B)\ ;\ B \subset (A \cup B)\ ;\ (A \cap B) \subset A\ ;\ (A \cap B) \subset B.$$

An examination of the Venn diagram of the union and intersection should make these results intuitively obvious.

Further, if A, B and C are three arbitrary sets, the following results are also true

$$A \cap (B \cap C) = (A \cap B) \cap C \qquad \ldots(2.6)$$

that is, the intersection is an *associative* operation.

$$A \cap (B \cup C) = (A \cap B) \cup (A \cap C) \qquad \ldots(2.7)$$

that is, the intersection operation is *distributive* over the union operation, and

$$A \cup (B \cap C) = (A \cup B) \cap (A \cup C) \qquad \ldots(2.8)$$

that is, the union operation is distributive over intersection operation.

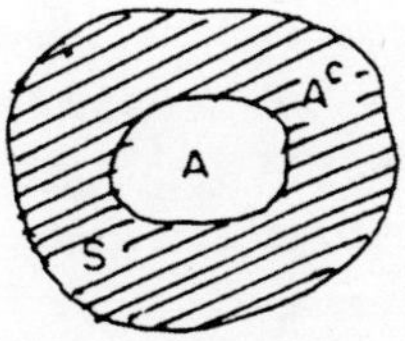

Fig. 2.8. Venn diagram representing the complement A^c.

The set of sample points of a sample space which are not in the set A is called the *complement* of set A and is denoted by A^c. The complement of the set A is represented by the shaded area in Fig. 2.8. The definition of the complement requires the specification of not only the set A whose complement is to be determined but also the sample space. With this definition for complement, we have

$$A \cup A^c = S \qquad \ldots(2.9)$$

$$A \cap A^c = 0 \qquad ...(2.10)$$

$$A = (A^c)^c \qquad ...(2.11)$$

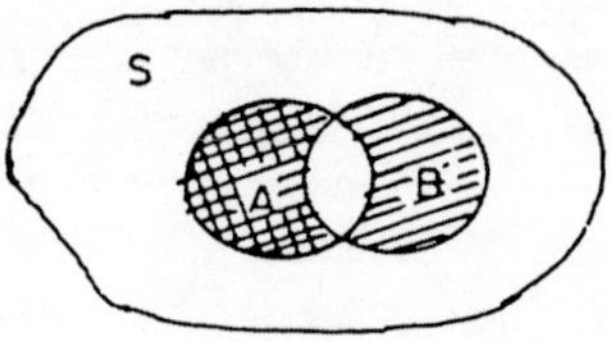

Fig. 2.9. Venn diagram of set differences
(*B*—*A*) and (*A*—*B*).

When we are not interested in all the points outside set *A*, but only in those outside points which are the elements of some other set *B*, we can define the *set difference* (*B*—*A*) of sets *A* and *B* to be the set of sample points of *B* which are not also points of *A*. Thus we see that the set difference (*B*—*A*) is, in effect, the complement of set *A* relative to set *B* rather than relative to the entire sample space *S*. Therefore, the set difference (*B*—*A*) is often called the *relative complement* of A with respect to *B*. The shaded portion of Venn diagram of Fig. 2.9 represents (*B*—*A*) while the double shaded portion represents (*A*—*B*), that is the relative complement of *B* with respect to *A*. With this definition we can write

$$A^c = (S - A) \qquad ...(2.12)$$

or

$$A = S - A^c \qquad ...(2.13)$$

Let *A* and *B* be two arbitrary sets in a sample space *S*. Then, according to *De Morgan's rule* called the principle of duality

$$(A \cap B)^c = A^c \cup B^c \qquad ...(2.14)$$

$$(A \cup B)^c = A^c \cap B^c \qquad ...(2.15)$$

The sets which have no elements in common are said to be *mutually exclusive* sets or *disjoint* sets and, therefore, if one occurs the other cannot occur. Hence if *A* and *B* are two mutually exclusive sets, it is obvious that their intersection is a null set. That is,

$$(A \cap B) = 0 \qquad ...(2.16)$$

If $A_1, A_2, \ldots\ldots, A_n$ are subsets of *S* such that they are mutually exclusive and collectively exhaustive, that is if

$$S = A_1 \cup A_2 \cup A_3 \cup \ldots\ldots \cup A_n = \bigcup_{i=1}^{n} A_i \qquad ...(2.17)$$

then the collection of sets A_1, A_2, ..., A_n is said to *partition* the sample space.

Example 2.1. *The annual peak discharge Q of a stream is believed not to exceed 3000 m³/s. Sketch an appropriate sample space* *to represent this and indicate the following events on the sample space.*

$$A = \{Q : 0 \le Q \le 1500\}\ ;\ B = \{Q : 1000 \le Q \le 2000\}.$$

Also indicate $A \cup B$ *and* $A \cap B$.

Sol. The sample points in this case can be represented on a line (one dimensional space) as shown in Fig. 2.10. The events A and B as defined in the problem, and also their union and intersection are indicated on the same sample space.

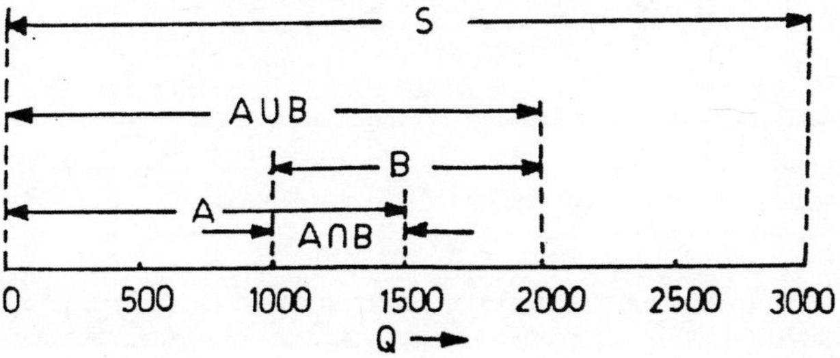

Fig. 2.10. Sample space and events of Example 2.1.

Example 2.2. *An experiment consists of observing the annual peak discharge of two streams just upstream of their confluence. It is believed that the annual peak discharge of one stream x can almost be 1500* m^3/s*, while the annual peak discharge of the other stream y cannot exceed 2500* m^3/s*. A project proposed just downstream of the confluence is designed for a peak flood of 2500* m^3/s*. Construct a suitable sample space for this experiment and on this sample space indicate the event that the project would not fail.*

Sol. The sample space for this experiment would be a two dimensional plane. Any point (x, y) on this plane is a sample point where the coordinates x and y denote the annual peak discharges in the two streams respectively. Fig. 2.11 shows this sample space. Since the project would not fail as long as the peak flood ≤ 2500 m^3/s, the region in the sample space for which the combined discharge indicated by any point is less than or equal to 2500 m^3/s represents the required event. Let this event be denoted by A. Then $A = \{x, y : 0 \le x + y \le 2500\}$. A line whose equation is $x + y = 2500$ is drawn in the sample space. All the points on this line or below this line will satisfy the condition that $x + y \le 2500$. Hence the dotted area in Fig. 2.11 represents the event A.

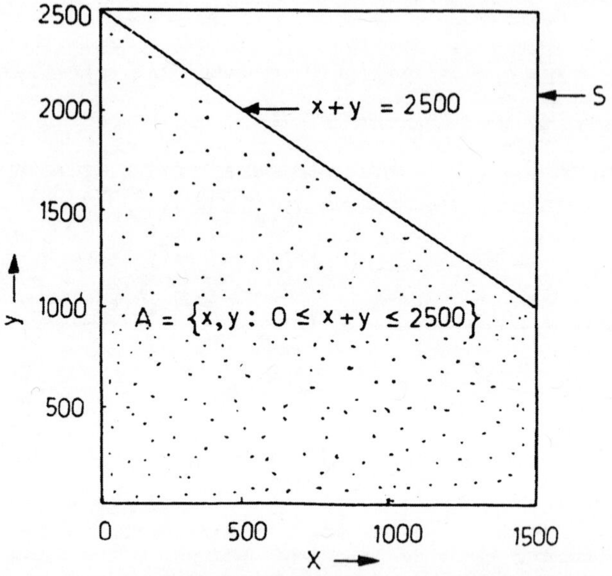

Fig. 2.11. Sample space and event of Example 2.2.

2.3. AXIOMS OF PROBABILITY

In the light of the concepts of sample points, sample space and the events or sets, we now discuss the axioms of probability.

Axiom 1. Given an experiment, there exists a sample space S representing the totality of possible outcomes of that experiment and a collection of subsets A of S called the events.

Axiom 2. To each event A in the collection of events, there can be assigned a non-negative real number $P[A]$, that is a number such that $0 \le P[A]$. This number is called the *probability* of event A.

Axiom 3. For any probability assignment P, it is required that $P[S] = 1$.

Axiom 4. If events A and B are mutually exclusive (that is, disjoint) events, then

$$P[A \cup B] = P[A] + P[B] \qquad ...(2.18)$$

Axiom 5. If events A_1, A_2, A_3, ... are mutually exclusive events, then

$$P[A_1 \cup A_2 \cup A_3 \cup ......] = P[A_1] + P[A_2] + P[A_3] + ...$$

or

$$P[\bigcup_{i=1}^{\infty} A_i] = \sum_{i=1}^{\infty} P[A_i] \qquad ...(2.19)$$

Axiom 1 says that outcomes of any experiment can be represented in the form of a sample space which can be divided into subsets by grouping the elements in the sample space. Axiom 2 says that the probability assigned to any of the events in the sample space is a real positive number. It does not specify how this probability can be assigned. The answer to this question can be determined only from a detailed knowledge of the physical situation under consideration. Axiom 3 says that the probability of the sure, or certain event is unity. It follows form axiom 3 that the probability of any event has an upper limit of 1 though it is not explicitly specified in Axiom 1. That is $0 \leq P[A] \leq 1$. Axiom 4 and Axiom 5 define probabilities of unions of events which are mutually exclusive.

2.4. INTERPRETATION OF PROBABILITY

The term probability will be interpreted in a number of ways. Three conceptual interpretations are now discussed.

The classical interpretation. If an event can occur in N equally likely and different ways, and if n of these events have an attribute A, then the probability of occurrence of A denoted by $P[A]$ is defined as n/N.

Thus, for example, in the case of a perfect die, the probability of rolling a face with even number is equal to 0.5, since there are six equally likely outcomes, that is, the six faces with numbers 1, 2, 3, 4, 5 and 6 of which 3 faces have the attribute that the face shown up is an even number. It is difficult to find hydrologic variables for which this interpretation can be meaningfully applied.

The frequency interpretation. If an experiment is conducted N times, and a particular attribute A occurs n times, then the limit of n/N as N becomes large is defined as the probability of the event A. $P[A]$.

The frequency interpretation is the one that is most generally used by modern statisticians and engineers. However, one may assert that even this definition is not general enough because it does not cover the case in which little or no experimental evidence exists.

The subjective interpretation. According to this interpretation, the probability $P[A]$ is a measure of the degree of belief one holds in a specified proposition A. Under this interpretation, probability may be directly related to the degree of belief one has in his judgement or prediction. The statement that the probability is 0.75 (or, equivalently that the odds are three to one) that the soil in the foundation is clay involves the subjective definition, since it represents our degree of belief concerning the foundation soil based on our experience. Such a statement would be meaningless under the first two interpretations. The classical interpretation is inadequate because there is no reason to believe that clay, sand, granite etc., are equally likely. The frequency interpretation is not applicable because we do not have historical data on what portion of times clay appears in the foundation. The subjective interpretation of probability is thus a broader concept than the classical and frequency interpretations and it includes the other two interpretations.

The basic rules of combining probabilities of events and many other results remain the same, irrespective of which of the foregoing interpretations is employed.

2.5. ELEMENTARY PROPERTIES OF PROBABILITY

The axioms of probability in conjunction with the concepts of set theory will lead to the following results.

(*i*) If A and B are two events that are not necessarily mutually exclusive, then

$$P[A \cup B] = P[A] + P[B] - P[A \cap B] \qquad ...(2.20)$$

Referring to Venn diagram of Fig. 2.6, $P[A] + P[B]$ counts the elements in the event $(A \cap B)$ twice. Since $(A \cup B)$ contains only the distinct elements of A and B, the elements of $(A \cap B)$ are to be deducted once from the above sum and hence the result.

(*ii*) If A is an event with probability $P[A]$, then

$$P[A^c] = 1 - P[A] \qquad ...(2.21)$$

From the algebra of sets, we have $S = A \cup A^c$ and also we know that A and A^c are mutually exclusive. Then from the Axioms 3 and 4, we get

$$P[S] = 1 \text{ and } P[S] = P[A \cup A^c] = P[A] + P[A^c].$$

Therefore $P[A] + P[A^c] = 1$, which proves Eq. (2.21).

(*iii*) If A and B are the events in a sample space S, then

$$P[B - A] = P[B] - P[A \cap B] \qquad ...(2.22)$$

(*iv*) If A and B are the events in a sample space S such that $A \subset B$, then

$$P[B - A] = P[B] - P[A] \qquad ...(2.23)$$

(*v*) If A, B and C are the events in a sample space S which are not necessarily mutually exclusive, then

$$P[A \cup B \cup C] = P[A] + P[B] + P[C] + P[A \cap B \cap C] - P[A \cap B] - P[B \cap C] - P[A \cap C] \qquad ...(2.24)$$

Example 2.3. *An engineer is designing a culvert across a stream. Based on the limited data available, the following probabilities have been assigned to the events of annual peak flow Q with the presumption that it cannot exceed 120 m^3/s.*

$$A = \{Q : 50 \le Q \le 100\}\ ;\ P[A] = 0.6$$
$$B = \{Q : 80 \le Q \le 120\}\ ;\ P[B] = 0.6$$
$$C = \{Q : 50 \le Q \le 120\}\ ;\ P[C] = 0.7$$

Describe the events $(A \cap B)$, A^c, $(B \cup A^c)$ and find the probabilities of these events.

Sol. The sample space and the other events are shown in Fig. 2.12 from which it is obvious that C is the union of A and B. That is, $C = (A \cup B)$.

$$P[C] = P[A \cup B] = 0.7$$
$$P[A \cup B] = P[A] + P[B] - P[A \cap B]$$
$$0.7 = 0.6 + 0.6 - P[A \cap B]$$
$$P[A \cap B] = 1.2 - 0.7 = 0.5$$
$$P[A^c] = 1 - P[A] = 1 - 0.6 = 0.4$$

From Fig. 2.12, we have

$$(B \cup A^c) = (S - C) \cup B$$

And since $(S - C)$ and B are mutually exclusive events

$$P[B \cup A^c] = P[S - C] + P[B]$$
$$= P[S] - P[C] + P[B]$$
$$= 1.0 - 0.7 + 0.6 = 0.9$$

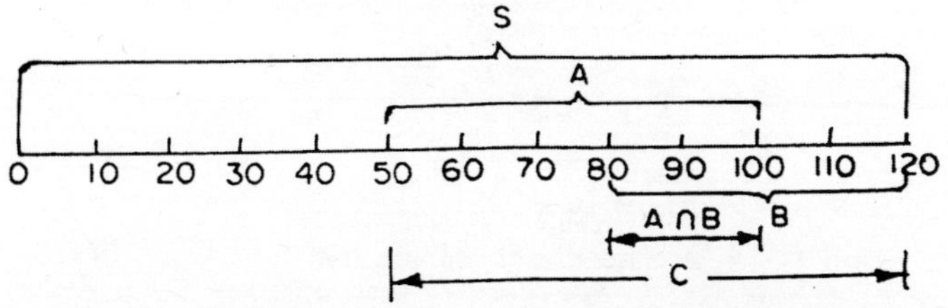

Fig. 2.12. Sample space and events of Example 2.3.

Therefore, the description of various events and their probabilities are as follows :

$$(A \cap B) = \{Q : 80 \le Q \le 100 \text{ m}^3/\text{s}\}\ ;\ P[A \cap B] = 0.5$$
$$A^c = \{Q : 0 \le Q \le 50 \text{ m}^3/\text{s or } 100 \le Q \le 120 \text{ m}^3/\text{s}\}\ ;\ P[A^c] = 0.4$$
$$(B \cup A^c) = \{Q : 0 \le Q \le 50 \text{ m}^3/\text{s or } 80 \le Q \le 120 \text{ m}^3/\text{s}\}\ ;$$
$$P[B \cup A^c] = 0.9.$$

Example 2.4. *Assuming that the probability is distributed uniformly over the sample space of example 2.2 (a highly questionable assumption indeed), determine the probability that the project would not fail.*

Sol. The probability that the project would not fail is same as the probability of the event A. Since the probability is uniformly distributed over the sample space, the probability of event A is equal to the ratio of area representing event A to the area representing the sample space. Referring to Fig. 2.11, we have

$$P[A] = \frac{(1500 \times 1000) + (\frac{1}{2} \times 1500 \times 1500)}{1500 \times 2500} = 0.9$$

Therefore there are ninety per cent chances that the project would not fail and only ten per cent chances that it would fail in any year.

2.6. CONDITIONAL PROBABILITY

The probability of an event A, given the knowledge that another event B has occurred is called the *conditional probability* and is denoted by $P[A/B]$. It is defined as the ratio of the probability of the intersection $(A \cap B)$ to the probability of event B.

$$P[A/B] = \frac{P[A \cap B]}{P[B]} \qquad \text{...(2.25)}$$

If $P[B]$ is zero, the conditional probability is undefined. Because, if $P[B]$ is zero, it implies that B is an empty set which does not belong to the sample space.

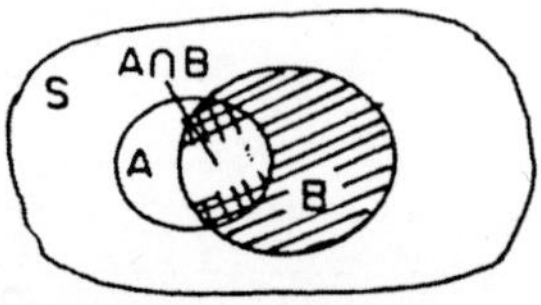

Fig. 2.13. Venn diagram for conditional probability.

To understand the concept of conditional probability consider the Venn diagram of Fig. 2.13. We are interested in the probability of A given the hypothesis that event B has already occurred. Once we insist upon the condition that event B occurs, we are no longer concerned with the probability that the sample point representing a particular outcome falls anywhere in the subset A of the sample space S. Rather, we are concerned with the probability that the sample point falls in that part of $(A \cap B)$ of A which is also a subset of points representing the event B. In essence then, instead of dealing with subset A of a sample space S, we must now deal with a subset $(A \cap B)$ of a new restricted sample space B, and our conditional probability of the event A must reflect this change in the effective sample spaces and hence the definition of conditional probability is as given by Eq. (2.25).

If two events are not related in any way and are independent of each other, the information that B has occurred should not affect the probability of event A. Therefore events A and B are said to be *independent* if and only if

$$P[A/B] = P[A] \quad ...(2.26)$$

From Eqs. (2.25) and (2.26), when two events A and B are independent, we can write

$$\frac{P[A \cap B]}{P[B]} = P[A]$$

or
$$P[A \cap B] = P[A] \,.\, P[B] \quad ...(2.27)$$

Either Eq. (2.26) or Eq. (2.27) can be used as a definition of independence. In general, events A, B, C, ... are mutually independent, if and only if

$$P[A \cap B \cap C \cap ......] = P[A] \,.\, P[B] \,.\, P[C] \quad ...(2.28)$$

Eq. (2.28) is a straight forward generalisation of Eq. (2.27). In other words, if events are mutually independent, the probability of their joint occurrence is simply equal to the product of their individual probabilities of occurrence.

Example 2.5. *If A and B are two events associated with a sample space and having probability P[A] = 0.4 and P[A ∪ B] = 0.7, what is P[B] when (i) A and B are mutually exclusive and (ii) A and B are mutually independent ?*

Sol. (*i*) When A and B are mutually exclusive, we have

$$P[A \cup B] = P[A] + P[B]$$
$$0.7 = 0.4 + P[B]$$
$$\therefore \quad P[B] = 0.3$$

(*ii*) When A and B are not mutually exclusive, we have

$$P[A \cup B] = P[A] + P[B] - P[A \cap B].$$

Since A and B are mutually independent,

$$P[A \cap B] = P[A] \,.\, P[B]. \text{ Thus,}$$
$$P[A \cup B] = P[A] + P[B] - P[A] \,.\, P[B]$$
$$0.7 = 0.4 + P[B] - 0.4 \,.\, P[B]$$
$$0.3 = P[B]\,(1 - 0.4) = 0.6\, P[B]$$

or
$$P[B] = 0.5.$$

Example 2.6. *A culvert is proposed just downstream of the confluence of two rivers A and B. Assume that the annual peak discharge from river A can take values of 0, 20, 40 and 60 m^3/s with probabilities of 0.1, 0.4, 0.4 and 0.1 respectively; and that the annual* *peak discharge from river B can take the values of 0, 10, 20 and 30 m^3/s with probabilities of 0.1, 0.4, 0.4 and 0.1 respectively. If a discharge of 50 m^3/s or more will halt the construction of the culvert find its probability. Assume that annual peak discharges from rivers A and B are independent. Is this assumption correct ?*

Sol. Denoting the annual peak discharges from rivers A and B by Q_A and Q_B, various events and their probabilities as given in the problem may be listed as given below :

$$A_1 = \{Q_A : Q_A = 0\}\,; \qquad P[A_1] = 0.1$$
$$A_2 = \{Q_A : Q_A = 20\}\,; \qquad P[A_2] = 0.4$$
$$A_3 = \{Q_A : Q_A = 40\}\,; \qquad P[A_3] = 0.4$$
$$A_4 = \{Q_A : Q_A = 60\}\,; \qquad P[A_4] = 0.1$$
$$B_1 = \{Q_B : Q_B = 0\}\,; \qquad P[B_1] = 0.1$$
$$B_2 = \{Q_B : Q_B = 10\}\,; \qquad P[B_2] = 0.4$$
$$B_3 = \{Q_B : Q_B = 20\}\,; \qquad P[B_3] = 0.4$$
$$B_4 = \{Q_B : Q_B = 30\}\,; \qquad P[B_4] = 0.1.$$

The discharge at the confluence $Q = Q_A + Q_B$ exceeds 50 m^3/s when any one of the events $\{A_2 \cap B_4)$, $(A_3 \cap B_2)$, $(A_3 \cap B_3)$, $(A_3 \cap B_4)$, $(A_4 \cap B_1)$, $(A_4 \cap B_2)$, $(A_4 \cap B_3)$ and $(A_4 \cap B_4)$ occurs. Therefore

$$P[Q \geq 50] = P[A_2 \cap B_4] + P[A_3 \cap B_2] + P[A_3 \cap B_3] + P[A_3 \cap B_4] + P[A_4 \cap B_1] + P[A_4 \cap B_2] + P[A_4 \cap B_3] + P[A_4 \cap B_4].$$

Since the discharges from the rivers A and B are given to be independent

$$P[Q \geq 50] = P[A_2]\,.\,P[B_4] + P[A_3]\,.\,P[B_2] + P[A_3]\,.\,P[B_3] + P[A_3]\,.\,P[B_4] + P[A_4]\,.\,P[B_1] + P[A_4]\,.\,P[B_2] + P[A_4]\,.\,P[B_3] + P[A_4]\,.\,P[B_4]$$
$$= 0.4 \times 0.1 + 0.4\,(0.4 + 0.4 + 0.1) + 0.1\,(0.1 + 0.4 + 0.4 + 0.1)$$
$$= 0.04 + 0.36 + 0.1 = 0.5$$

Hence the probability that construction work at the culvert is halted in any year is 0 5.

The assumption that the discharges from the two rivers are independent is not justifiable. Since the catchment areas of the two streams are adjacent and fairly small (as evident from the magnitudes of annual peak flows), the storm which produces annual peak discharge on one stream is also likely to produce annual peak discharge on the other stream. Therefore we can expect some strong correlation between the peak discharges and they cannot be treated, strictly speaking, as independent. However, the example is presented here only to provide an understanding of the concept of conditional probability and independence.

Example 2.7. *The observed annual flood peaks of a stream for a period of 40 years from 1951 to 1990 in m^3/s are given below :*

1951—60 : 395, 619, 766, 422, 282, 990, 705, 528, 520, 436
1961—70 : 697, 624, 496, 589, 598, 359, 686, 726, 527, 310
1971—80 : 408, 721, 814, 459, 440, 632, 343, 634, 464, 373
1981—90 : 289, 371, 522, 342, 446, 366, 699, 560, 450, 610.

Determine the probability that an annual peak discharge in excess of 700 m^3/s will occur in 2 successive years on the stream.

Sol. It can be seen from the given data that a peak flow in excess of 700 m^3/s occurred 6 times (in 1953, 56, 57, 68, 72 and 73) in the 40 years record. Hence from frequency interpretation the probability of the event that the annual peak flow exceeds 700 m^3/s in any year is 6/40, that is, 0.15. If it is assumed that the peak flows from year to year are independent, then applying Eq. (2.27), the probability of peak flow exceeding 700 m^3/s in two successive years would be $0.15 \times 0.15 = 0.0225$.

Example 2.8. *A study of daily rainfall record at a certain raingauge station has revealed that in the month of August the probability of a rainy day following a rainy day is 0.45, a dry day following a dry day is 0.75, a rainy day following a dry day is 0.275 and a dry day following a rainy day is 0.555. If it is observed that a certain day in August is a rainy day, what is the probability that (i) the next two days will also be rainy days ? (ii) the next two days will be dry days ?*

Sol. (*i*) Following some initial rainy day, let A be the rainy day 1 and B be the rainy day 2. Now from the given information $P[A] = 0.45$ since A happens to be a rainy day following the initial rainy day. Now consider the probability of the event that second day is also a rainy day given that first day is also a rainy day. That is, consider $P[B/A]$. Now

$$P[B/A] = \frac{P[A \cap B]}{P[A]}; \quad \text{or} \quad P[A \cap \mathrm{B}] = P[A]\, P[B/A].$$

But $P[B/A]$ is also 0.45 since this is the probability of rainy day following rainy day.

$$P[A \cap B] = 0.45 \times 0.45 = 0.2025.$$

(*ii*) Following some initial rainy day let A be the dry day 1 and B be the dry day 2. From the given information $P[A] = 0.555$ since this is a dry day following a rainy day. Now

$$P[A \cap B] = P[A]\, P[B/A]$$

But $P[B/A]$ now is 0.75 since this is the probability of a dry day following a dry day. Therefore

$$P[A \cap B] = 0.555 \times 0.75 = 0.41625.$$

In other words, given an initial rainy day in August, the probability that the next two days are rainy days is 0.2025 and the probability that the next two days are dry days is 0.41625.

2.7. TOTAL PROBABILITY AND BAYES' THEOREM

Let $A_1, A_2, \ldots\ldots, A_n$ represent a set of mutually exclusive and collectively exhaustive events which partition the sample space S. Let B be any arbitrary event in the same sample space as shown in Fig. 2.14. Then according to the *total probability theorem*

$$P[B] = P[B/A_1]\,P[A_1] + P[B/A_2]\,P[A_2] + \ldots\ldots + P[B/A_n]\,P[A_n]$$

or
$$P[B] = \sum_{i=1}^{n} P[B/A_i]\,P[A_i] \qquad \ldots(2.29)$$

This theorem can be easily proved as follows. From the Venn diagram of Fig. 2.14, we have $B = (B \cap A_1) \cup (B \cap A_2) \cup \ldots\ldots \cup (B \cap A_n)$.

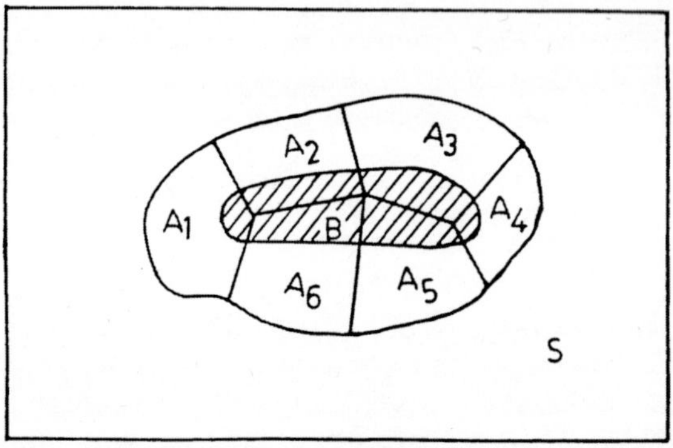

Fig. 2.14. Venn diagram for theorem of total probability.

Since $A_1, A_2, \ldots\ldots, A_n$ are mutually exclusive, the events $(B \cap A_1)$, $(B \cap A_2)$, ..., $(B \cap A_n)$ are also mutually exclusive. Hence

$$P[B] = P[B \cap A_1] + P[B \cap A_2] + \ldots\ldots + P[B \cap A_n] \qquad \ldots(2.30)$$

And from the definition of conditional probability

$$P[B \cap A_i] = P[B/A_i]\,P[A_i] \qquad \ldots(2.31)$$

Substitution of Eq. (2.31) in Eq. (2.30) yields Eq. (2.29) and thus the total probability theorem is proved. From the definition of conditional probability, we also have that

$$P[A_i/B] = P[A_i \cap B]/P[B]$$

or
$$P[B \cap A_i] = P[A_i/B]\,P[B] \qquad \ldots(2.32)$$

From Eq. (2.31) and Eq. (2.32), we can write

$$P[B/A_i]\,P[A_i] = P[A_i/B]\,P[B]$$

$$P[A_i/B] = P[B/A_i]\,P[A_i]/P[B] \qquad \text{...(2.33)}$$

Substituting Eq. (2.29) for $P[B]$ in Eq. (2.33), we get

$$P[A_i/B] = \frac{P[B/A_i]\,P[A_i]}{\sum_{j=1}^{n} P[B/A_j]\,P[A_j]} \qquad \text{...(2.34)}$$

Eq. (2.34) is called the *Bayes' theorem* or *Bayes' rule*. The terms *a priori* and *a posteriori* are often used for the probabilities $P[A_i]$ and $P[A_i/B]$. Eq. (2.34) is the basis for the *Bayesian Decision theory*. Bayes' theorem provides a method of estimating probability of one event by observing a second event.

2.8. RANDOM VARIABLE

In the preceeding sections we have seen that in using the probability theory, we need to represent the sample space and subgrouping of the elements in sample space as subsets or events. It may be tedious to describe the sample space if the elements of the sample space are not numbers. We can eliminate this difficulty by representing the elements of the sample space by numbers x, or ordered set of numbers $(x_1, x_2, \ldots\ldots, x_n)$. For example, consider the experiment of tossing a coin. The sample space associated with this experiment is $S = \{c : \text{where } c \text{ is } T \text{ or } H\}$, where T and H represent the tails and heads respectively. Let X be a function such that $X(c) = 0$ if c is T and $X(c) = 1$ if c is H. Thus X is a single valued, real valued function defined on the sample space which takes us from the sample space to space of real numbers. If the outcome of the experiment itself is a real number, the problem is very much simplified.

Given a random experiment with sample space S, a function X which assigns to each element C in the sample space one and only one real number $X(c) = x$ is called a *random variable*. Now the space of X is the set of real numbers.

In the experiment of flipping a coin, we can write $P[X = 0]$ $= P$[outcome is Head] $= P[c, c \text{ is } H]$. Similarly we can write $P[X = 1]$ $= P$ [outcome is Tail] $= P[c, c \text{ is } T]$. Thus, we see that a random variable which carries the probability from a sample space to a space of real numbers. The concept of a random variable enables us to map the qualitative variables on to a quantitative scale.

A sample space involving either a finite number or a countably infinite number of elements is said to be discrete. Any random variable which is defined on such a sample space can also take on a finite number or countably infinite number of values. Such a variable is called a *discrete random variable*. The examples of discrete random variables in hydrology are the numbers of storms in a given month, the number of rainy days in a year, the number of floods in a year causing a peak flow above a stipulated discharge etc.

A random variable is said to be a *continuous random variable* if it can take any value in a given interval or intervals. Continuous random variables are associated with random phenomena whose outcome has to be measured rather than counted. The discharge in a river, the outflow from a lake, the depth to ground water table, the annual rainfall at a location, the direction of wind etc., are some of the continuous random variables that we come across in hydrology.

If the space of real numbers of a random variable consists of some isolated points as well as some continuous intervals, then it is called a *mixed random variable*.

A function of a random variable is also a random variable. If X is a random variable then $Z = g(X)$ is a random variable as well.

A random variable is usually denoted by an upper case character such as X, while any value the random variable takes in an experiment is denoted by the corresponding lower case character such as x. We shall now discuss two methods of describing the probabilities of random variables namely (*i*) *probability mass function* and *probability density function* and (*ii*) *distribution function* or also known as *cumulative distribution function*.

2.9. PROBABILITY FUNCTION AND DISTRIBUTION FUNCTION OF A DISCRETE RANDOM VARIABLE

The probability function, also known as probability mass function (abbreviated as p.m.f.) of a discrete random variable X is an expression $p_X(x_i)$ that gives the probability associated with all possible values of the random variable. That is $p_X(x_i) = P[X = x_i]$, $i = 1, 2, ...$ The suffix X is only to indicate that the probability function pertains to the random variable X. As there is no fear of anbiguity, the suffix will be omitted hereafter. That is

$$p(x_i) = P[X = x_i] \qquad ...(2.35)$$

To satisfy the axioms of probability, the probability functions must clearly fulfil the following conditions.

$$0 \leq p(x_i) \leq 1 \qquad ...(2.36)$$

$$\sum_{\text{all } x_i} p(x_i) = 1 \qquad \text{...(2.37)}$$

$$P[a \le X \le b] = \sum_{x_i \le a}^{x_i \le b} p(x_i) \qquad \text{...(2.38)}$$

The summation in the above expressions are, of course, over those values of x only where the probability function is defined.

An equivalent way of describing the behaviour of a discrete random variable is through the use of a distribution function, also known as cumulative distribution function (abbreviated as c.d.f.). The value of the distribution function $F_X(x)$ is simply the probability of the event that the random variable takes on any value equal to or less than x. Again the suffix X will be omitted. That is

$$F(x) = P[X \le x] \qquad \text{...(2.39)}$$

In terms of the probability function, this can be written as

$$F(x) = \sum_{\text{all } x_i \le x} p(x_i) \qquad \text{...(2.40)}$$

The distribution function has jumps in it at each x_i which are equal to $p(x_i)$ in magnitude. Using the distribution function, the probability that $X = x_i$ can be determined from the equation

$$P[X = x_i] = F[x_i] - F(x_{i-1}) \qquad \text{...(2.41)}$$

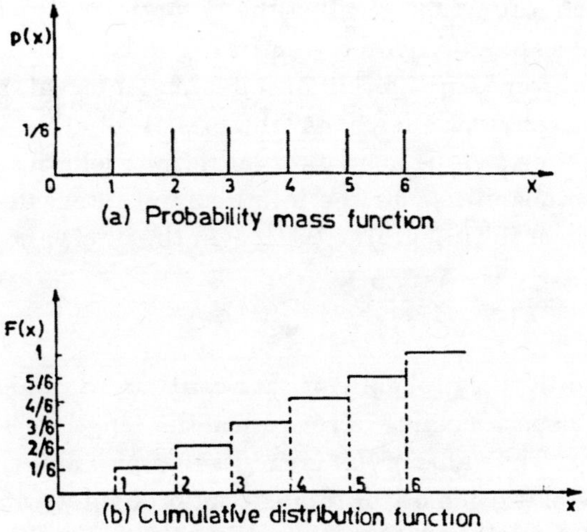

Fig. 2.15. Probability representation of die tossing experiment.

Let us consider the experiment of tossing a die and let us define the random variable as the number of the face that shows up. The random variable under consideration is of discrete type which can take any of the values 1, 2, 3, 4, 5 and 6. If it is a fair die, every face has equal chance to show up. Thus, the probability function is given by $p(x_i) = 1/6$ for $x_i = 1, 2, 3, 4, 5$ and 6. This is shown in Fig. 2.15 (*a*). The distribution function derived from this proba-bility function is shown in Fig. 2.15 (*b*). As seen from the figure, the distribution function has constant ordinate between the jumps which occur at those values where the probability function is defined.

2.10. PROBABILITY DENSITY FUNCTION AND DISTRIBUTION FUNCTION OF A CONTINUOUS RANDOM VARIABLE

In the case of a continuous random variable we cannot assign probabilities for every value that it can take. This is because it can take, theoretically, infinite number of values in its range or interval. However, the problem of specifying the probability distribution of a continuous random variable X can easily be managed. If the x-axis is divided into a large number of mutually exclusive intervals each of an infinitesimal length dx, we can think of a function $f_X(x)$ such that the probability that X lies in the interval x to $x + dx$ is given by $f_X(x)\, dx$. Such a function is called the *probability density function* (abbreviated as p.d.f.). Again the suffix X will be omitted hereafter. Since the occurrence in different infinitesimal intervals are mutually exclusive events, it follows that the probability that the random variable takes on a value in an interval of finite length x_1 to x_2 is the sum of the probabilities for the infinitesimal intervals within the finite interval or the integral of $f(x)\, dx$ over the interval x_1 to x_2. Thus

$$P[x_1 \le X \le x_2] = \int_{x_1}^{x_2} f(x)\, dx \qquad \text{...(2.42)}$$

Evidently, the probability that a continuous random variable X takes on a specific value is zero since the length of the interval has vanished. The value of $f(x)$ is not itself a probability ; it is only the measure of the density or intensity of probability at that point. Therefore $f(x)$ need not be less than 1 but the two conditions given below must be satisfied by it.

$$f(x) \ge 0 \qquad \text{...(2.43)}$$

$$\int_{-\infty}^{\infty} f(x)\, dx = 1 \qquad \text{...(2.44)}$$

Since the probability for any interval cannot be negative, $f(x)\,dx$ cannot be negative for any dx. Hence Eq. (2.43). Similary the probability that the random variable takes on any value in its range is a sure event. Hence Eq. (2.44) must be satisfied.

As in the case of discrete variable, the cumulative distribution function of a continuous random variable is an alternative form of describing its probability law. Its definition remains the same as in the case of discrete random variable. That is

$$F(x) = P[X \leq x] \qquad \text{[Eq. (2.39)]}$$

and without loss of generality the right hand side of Eq. (2.39) may be written as $P[-\infty \leq X \leq x]$. Making use of Eq. (2.42) one can write

$$F(x) = P[-\infty \leq X \leq x] = \int_{-\infty}^{x} f(u)\, du \qquad \text{...(2.45)}$$

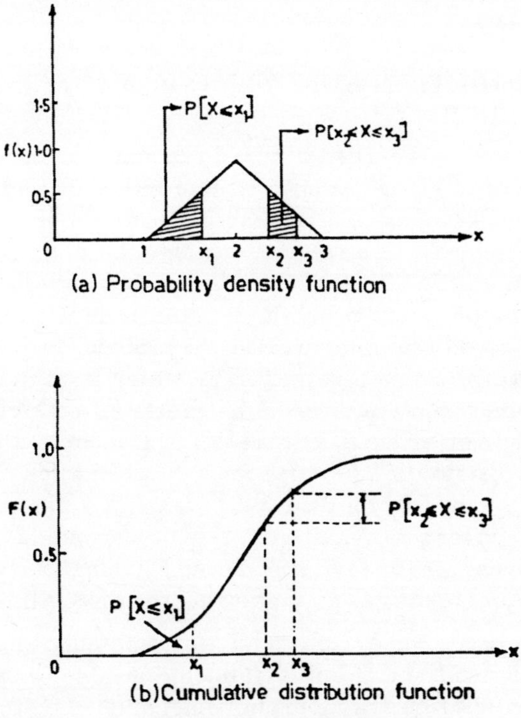

Fig. 2.16. Probability representation of a continuous random variable.

where u has been used as a dummy variable of integration to avoid confusion with the limit of integration x. It is apparent that the probability that X takes on a value between x_1 and x_2 is given by

$$P[x_1 \le X \le x_2] = \int_{x_1}^{x_2} f(x)\, dx \qquad ...(2.46)$$

or

$$P[x_1 \le X \le x_2] = F(x_2) - F(x_1) \qquad ...(2.47)$$

Since the probability that a continuous random variable takes on a specific value is zero, the expressions $P[x_1 \le X \le x_2]$, $P[x_1 < X \le x_2]$, $P[x_1 \le X < x_2]$ and $P[x_1 < X < x_2]$ are all equivalent.

The probability density function and the corresponding cumulative distribution function of a continuous random variable having a triangular distribution are shown in Fig. 2.16 (*a*) and Fig. 2.16 (*b*) respectively.

If $f(x)$ and $F(x)$ are, respectively, probability density function and cumulative distribution function, then $f(x) = \dfrac{dF(x)}{dx}$, where the derivative exists.

2.11. DISTRIBUTION FUNCTION OF A MIXED RANDOM VARIABLE

As pointed out earlier, a mixed random variable can take on infinite number of values in one or more intervals and it can also take on one or more specific values. For example, if we define the daily rainfall amount as a random variable there is a finite probability that the rainfall is zero while the probability distribution of daily rainfall amount on rainy days would form a continuous distribution. A second example would be the probability distribution of the water level in some reservoir. The water level may be maintained at some constant level d as much as possible but may fluctuate below or above d at times. Fig. 2.17 shows the density function and distribution function for such a situation. The density function is given by $f_1(x)$ for $0 \le x < d$, $p(d) = P[X = d]$ represented as a spike and $f_2(x)$ for $x > d$. Similarly, the corresponding distribution function is given by $F_1(x)$ for $0 \le x < d$, and $F_2(x)$ for $x > d$ with a jump of a height equal to $p(d)$ at $x = d$ showing the discontinuity.

In the discussion so far, we have attempted to show that it is convenient to use the concept of random variable (instead of sets and events) to describe a random phenomenon. Associated with each random variable is a distribution which, when known, offers the complete description of the probabilistic behaviour of the underlying

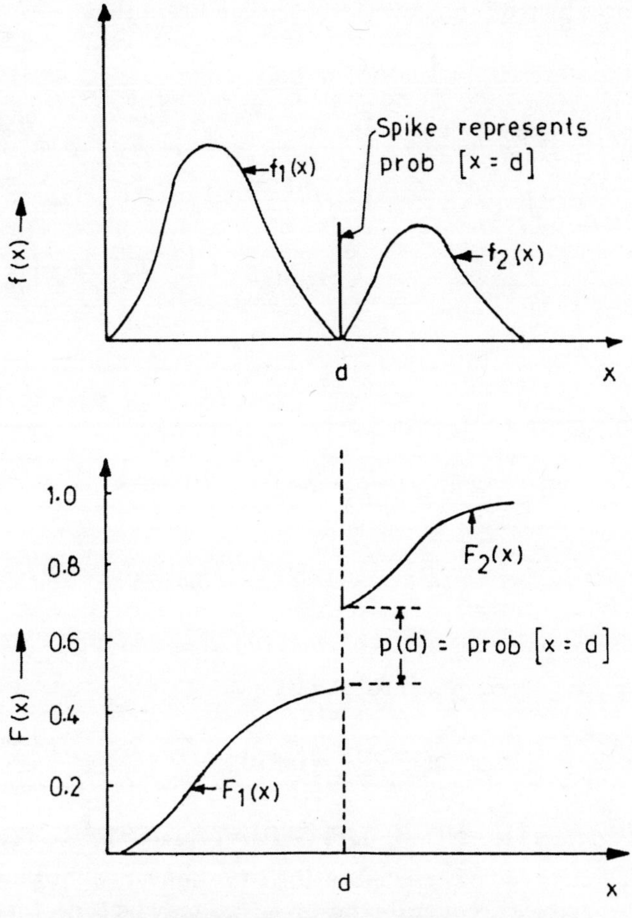

Fig. 2.17. Density and distribution functions of a mixed random variable.

physical situation. The observed data on the random variable forms the basis for identifying proper distribution and deriving it.

Example 2.9. *The duration of a monsoon storm X is a random variable whose probability density function is given by*

$$f(x) = kx^2 \quad \text{for } 0 \le x \le 6 \text{ h}$$
$$= k(12 - x)^2 \text{ for } 6 \le x \le 12 \text{ h}$$
$$= 0 \quad \text{elsewhere}$$

Evaluate the constant k. What is the probability that the duration of the storm is between 6 h and 9 h ? What is the probability that the duration of the storm is between 6 h and 9 h, given that the storm had lasted for 6 h ?

Sol. The p.d.f. has to satisfy the condition that

$$\int_{-\infty}^{\infty} f(x)\, dx = 1.$$

In the present case, we have

$$\int_{-\infty}^{\infty} f(x)\, dx = \int_{-\infty}^{0} f(x)\, dx + \int_{0}^{6} f(x)\, dx + \int_{6}^{12} f(x)\, dx + \int_{12}^{\infty} f(x)\, dx$$

$$1 = 0 + \int_{0}^{6} kx^2\, dx + \int_{6}^{12} (12 - x)^2\, dx + 0$$

This gives $k = 1/144$.

(*i*) $P[6 \le X \le 9] = \int_{6}^{9} f(x)\, dx$

$$= \int_{6}^{9} \frac{(12 - x)^2}{144}\, dx = 0.4375$$

(*ii*) From the definition of conditional probability

$$P[6 \le X \le 9/X \ge 6] = \frac{P[6 \le X \le 9]}{P[X \ge 6]}$$

$$= \frac{0.4375}{0.5} = 0.875.$$

2.12. BIVARIATE AND MULTIVARIATE DISTRIBUTIONS

So far we have been discussing situations involving only one random variable. Frequently, however, we may be concerned with situations wherein the simultaneous occurrence of two or more random variables is of interest ; for example, the flow rates of two streams near their confluence. We shall now consider mainly the bivariate case involving two random variables and the multivariate case follows as a straight forward generalisation of this.

If X and Y are discrete random variables, the function given by

$$p(x, y) = P[X = x, Y = y] \qquad ...(2.48)$$

for each pair of values of (x, y) within the range of X and Y is called the *joint probability function* of X and Y. The joint probability function must satisfy the following conditions :

(*a*) $p(x, y) \ge 0$ for each pair of values (x, y) within its domain.

(b) $\sum_x \sum_y p(x, y) = 1$, where the double summation extends over all possible pairs (x, y) within its domain.

The *joint cumulative distribution function* is given by

$$F(x, y) = P[X \leq x, Y \leq y] = \sum_{s \leq x} \sum_{t \leq y} p(s, t) \qquad ...(2.49)$$

for $-\infty < x < \infty$, $-\infty < y < \infty$, where $p(s, t)$ is the value of the joint probability function of X and Y at (s, t).

Two continuous random variables X and Y are said to have the *bivariate or joint probability density function* $f(x, y)$, if

$$P[x_1 \leq X \leq x_2, y_1 \leq Y \leq y_2] = \int_{x_1}^{x_2} \int_{y_1}^{y_2} f(x, y)\, dx\, dy \qquad ...(2.50)$$

The joint probability density function has to satisfy the following conditions.

(a) $f(x, y) \geq 0$, for $-\infty < x < \infty$, $-\infty < y < \infty$

(b) $\int_{-\infty}^{\infty} \int_{-\infty}^{\infty} f(x, y)\, dx\, dy = 1$.

The *bivariate or joint cumulative distribution function* is defined to be

$$F(x_1, y_1) = P[-\infty < X \leq x_1, -\infty < Y \leq y_1]$$

$$= \int_{-\infty}^{x_1} \int_{-\infty}^{y_1} f(s, t)\, ds\, dt \qquad ...(2.51)$$

for $-\infty < x_1 < \infty$, $-\infty < y_1 < \infty$, where $f(s, t)$ is the value of the joint probability density of X and Y at (s, t).

Analogous to the relationship $f(x) = \frac{dF(x)}{dx}$ in the case of single random variable, for the bivariate case the relationship that $f(x, y) = \frac{\partial^2 F(x, y)}{\partial x \, \partial y}$ holds wherever these partial derivatives exist.

Similar definitions apply for a multivariate case involving three or more (but finite number) random variables.

Example 2.10. *The joint probability density function of two random variables X and Y is given by*

$$f(x, y) = e^{-(x+y)} \quad \text{for } x > 0, y > 0$$
$$= 0 \quad \text{elsewhere.}$$

Find the probability that $(1 < X < 3$, *and* $1 < Y < 2)$.

Sol. $$P[1 < X < 3, 1 < Y < 2] = \int_1^2 \int_1^3 f(x, y)\, dx\, dy$$

$$= \int_1^2 \int_1^3 e^{-(x+y)}\, dx\, dy = 0.074.$$

2.13. MARGINAL DISTRIBUTIONS

Sometimes we are given the joint distribution of two random variables and we would like to obtain the distribution of one of the variables alone. This can be obtained by integrating out the variable that is no longer of interest. Then we will get the *marginal distribution* of the other variable. Thus if $p(x, y)$ is the joint probability function of the two discrete random variables X and Y, $g(x)$ is the probability mass function of X, and $h(y)$ is the probability mass function of Y, then

$$g(x) = \sum_{\text{all } y} p(x, y) \qquad ...(2.52)$$

for each value of x within the range of X, and

$$h(y) = \sum_{\text{all } x} p(x, y) \qquad ...(2.53)$$

for each value of y within the range of Y.

Similarly if X and Y are two continuous random variables with joint probability density function $f(x, y)$ and if $g(x)$ is the probability density function of X and $h(y)$ is the probability density function of Y, then

$$g(x) = \int_{-\infty}^{\infty} f(x, y)\, dy \quad \text{for } -\infty < x < \infty \qquad ...(2.54)$$

$$h(y) = \int_{-\infty}^{\infty} f(x, y)\, dx \quad \text{for } -\infty < y < \infty \qquad ...(2.55)$$

When we are dealing with more than two variables, we can speak of not only the marginal distributions of the individual random variables, but also of the *joint marginal distributions* of several of random variables. For example, if the joint probability density function of the continuous random variables $X_1, X_2, ..., X_n$ is $f(x_1, x_2, ..., x_n)$, then the marginal distribution of X_1 alone is given by

$$g(x_1) = \int_{-\infty}^{\infty} ...... \int_{-\infty}^{\infty} f(x_1, x_2, ..., x_n)\, dx_2\, dx_3 ...... dx_n \qquad ...(2.56)$$

Whereas the joint marginal distribution of X_1, X_2 and X_3 is given by

$$h(x_1, x_2, x_3) = \int_{-\infty}^{\infty} \ldots\ldots \int_{-\infty}^{\infty} f(x_1, x_2, \ldots\ldots, x_n)\, dx_4\, dx_5 \ldots\ldots dx_n \qquad \ldots(2.57)$$

Example 2.11. *Given the joint probability density function*

$$f(x, y) = \tfrac{2}{3}(x + 2y) \text{ for } 0 < x < 1,\ 0 < y < 1$$
$$= 0 \qquad \text{elsewhere}$$

find the marginal functions of X and Y.

Sol. Performing the necessary integrations

$$g(x) = \int_{-\infty}^{\infty} f(x, y)\, dy = \int_0^1 \frac{2}{3}(x + 2y)\, dy$$

$$g(x) = \tfrac{2}{3}(x + 1) \quad \text{for } 0 < x < 1$$
$$= 0 \qquad \text{elsewhere}$$

Similarly $$h(y) = \int_{-\infty}^{\infty} f(x, y)\, dx = \int_0^1 \frac{2}{3}(x + 2y)\, dx$$

$$h(y) = \tfrac{1}{3}(1 + 4y) \text{ for } 0 < y < 1$$
$$= 0 \qquad \text{elsewhere.}$$

2.14. CONDITIONAL DISTRIBUTIONS

Earlier in Section 2.6, we defined the conditional probability of event A given that event B has occurred as $P[A / B] = P[A \cap B]/P[B]$ provided $P[B] \neq 0$. This definition of random events can be extended to random variables also. If $f(x, y)$ is the joint probability density function of the two continuous random variables X and Y, $h(y)$ is the marginal density of Y, then the conditional density of X is given by

$$f(x/y) = \frac{f(x, y)}{h(y)} \qquad \ldots(2.58)$$

provided $h(y) \neq 0$. Similarly if $g(x)$ is the marginal density of X, then the conditional density of Y is given by

$$f(y/x) = \frac{f(x, y)}{g(x)} \qquad \ldots(2.59)$$

provided $g(x) \neq 0$.

When we are dealing with more than two random variables, we can consider various different kinds of conditional densities. For instance, if $f(x_1, x_2, x_3, x_4)$ is the joint density function of the continuous random variables X_1, X_2, X_3 and X_4, we can write

$$f(x_3/x_1, x_2, x_4) = \frac{f(x_1, x_2, x_3, x_4)}{g(x_1, x_2, x_4)} \quad ...(2.60)$$

where $g(x_1, x_2, x_4) \neq 0$ is the joint marginal density of X_1, X_2 and X_4. Likewise

$$f(x_2, x_4/x_1, x_3) = \frac{f(x_1, x_2, x_3, x_4)}{h(x_1, x_3)} \quad ...(2.61)$$

where $h(x_1, x_3) \neq 0$ is the joint marginal density of X_1 and X_3.

Expressions similar to Eq. (2.58) through Eq. (2.61) can be defined when we are dealing with discrete random variables.

Questions of *independence* are usually of great importance when we are dealing with two or more random variables. If the conditional density of X is independent of Y, then we have $f(x/y) = g(x)$ and hence from Eq. (2.58) we can write

$$f(x/y) = g(x) = f(x, y)/h(y)$$

or
$$f(x, y) = g(x) \cdot h(y) \quad ...(2.62)$$

That is, when two random variables are independent their joint probability density function is simply the product of the individual marginal density functions of the variables. Generalising this, the random variables $X_1, X_2, ..., X_n$ are said to be independent if and only if

$$f(x_1, x_2, ..., x_n) = f_1(x_1)\, f_2(x_2) \,......\, f_n(x_n) \quad ...(2.63)$$

for all $(x_1, x_2, ..., x_n)$ within their range.

Example 2.12. *Are the random variables of Example 2.11 independent?*

Sol. We have $f(x, y) = \frac{2}{3}(x + 2y), g(x) = \frac{2}{3}(x + 1)$

and $h(y) = \frac{1}{3}(1 + 4y)$.

Now $$f(x/y) = \frac{f(x, y)}{h(y)} = \frac{\frac{2}{3}(x + 2y)}{\frac{1}{3}(1 + 4y)} = \frac{2x + 4y}{1 + 4y}.$$

Thus we see that $f(x/y) \neq g(x)$. Similarly it can be observed that $f(y/x) \neq h(y)$. Hence the random variables X and Y are not independent.

Example 2.13. *Two independent random variables X and Y have the probability density functions given by*

$$g(x) = 2\,a\,e^{-bx} \quad \text{for } 0 \le x \le \infty$$

$$h(y) = 2\,a\,e^{-by} \quad \text{for } 0 \le y \le \infty$$

Find the probability that value of X exceeds value of Y in any observation.

Sol. Since the random variables are independent, we have

$$f(x, y) = g(x) \,.\, h(y) = 4a^2 e^{-b(x+y)}$$

Now $P[0 \le X < \infty, 0 \le Y \le x] = \int_{y=0}^{x} \int_{x=0}^{\infty} f(x, y)\, dx\, dy$

$$= \int_0^x \int_0^\infty 4a^2 e^{-b(x+y)} \,.\, dx\, dy = 3a^2/b^2.$$

2.15. DERIVED DISTRIBUTIONS

As pointed out earlier, any function of a random variable is also a random variable. Here we will present a method of determining the probability distribution of a functionally dependent random variable when the probability density function of the independent variable is known.

Let X be a random variable whose probability density function $f(x)$ is known and let Y be another random variable related to X through some functional relationship, say, $Y = h(X)$. Let $g(y)$ be the probability density function of Y which is to be determined. For simplicity we assume that the relationship between X and Y is one to one. This means for any value of X, there is one and only one value of Y. [Methods to find $g(y)$ even when this assumption does not hold are also available, but not discussed here.] Referring to Fig. 2.18, we see that the probability that X falls in the internal dx is same as the probability that Y falls in the corresponding internal dy. This is because whenever X falls in dx, Y automatically falls in dy. Hence we can write

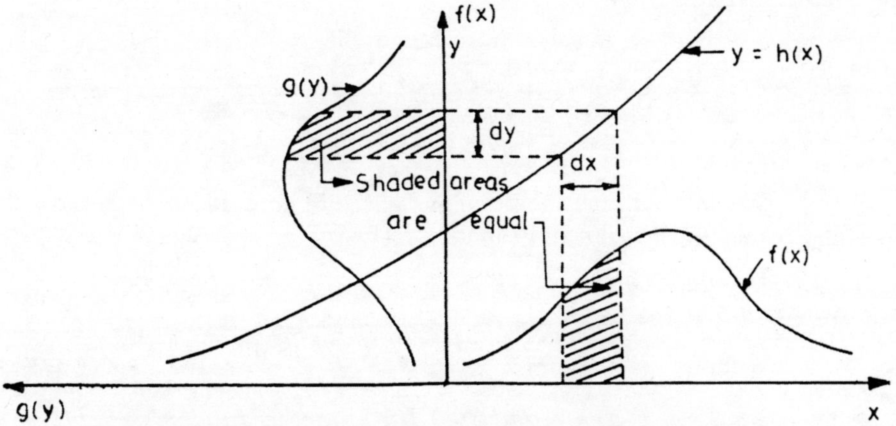

Fig. 2.18. Graphical representation for derived distribution.

$$f(x)\,dx = g(y)\,dy$$

or
$$g(y) = f(x)\,\frac{dx}{dy}$$

Since $g(y)$ cannot be negative, we specify that

$$g(y) = f(x)\,.\left|\frac{dx}{dy}\right| \qquad \text{...(2.64)}$$

The functional relationship $Y = h(X)$ can be uniquely solved for x to give $x = w(y)$, which may be called the *inverse transformation*. In Eq. (2.64) wherever x appears it will be replaced by $w(y)$ to yield $g(y)$ completely in terms of y. The probability density function thus obtained is called a *derived distribution.*

This technique can be extended to the cases of two or more random variables also. Let $f(x_1, x_2)$ be the joint probability density function of the two continuous random variables X_1 and X_2. If the functions given by $y_1 = u_1\,(x_1, x_2)$ and $y_2 = u_2\,(x_1, x_2)$ are partially differentiable with respect to both x_1 and x_2 and represent a one-to-one tranformation for all values within the ranges of X_1 and X_2 for which $f(x_1, x_2) \neq 0$, then for these values of x_1 and x_2 the equations $y_1 = u_1\,(x_1, x_2)$ and $y_2 = u_2\,(x_1, x_2)$ can be unequally solved for x_1 and x_2 to give $x_1 = w_1(y_1, y_2)$ and $x_2 = w_2(y_1, y_2)$ and for the corresponding values of y_1 and y_2 the joint probability densities of $Y_1 = u_1\,(X_1, X_2)$ and $Y_2 = u_2\,(X_1, X_2)$ is given by

$$g(y_1, y_2) = f(x_1, x_2)\,.\,|\,J\,|$$

or
$$g(y_1, y_2) = f[w_1(y_1, y_2), w_2(y_1, y_2)]\,.\,|\,J\,| \qquad \text{...(2.65)}$$

Here J is called the Jacobian of the transformation. It is the determinant given by

$$J = \begin{vmatrix} \dfrac{\partial x_1}{\partial y_1} & \dfrac{\partial x_1}{\partial y_2} \\ \dfrac{\partial x_2}{\partial y_1} & \dfrac{\partial x_2}{\partial y_2} \end{vmatrix} \qquad \text{...(2.66)}$$

We will not give the proof of Eq. (2.65) here. It can be found in advanced textbooks on Probability Theory.

Example 2.14. *The random variable X has the probability density function given by*

$$f(x) = \frac{1}{\sigma\sqrt{2\pi}}\,e^{-(x-\mu)^2/2\sigma^2}$$

for $-\infty < x < \infty$, $-\infty < \mu < \infty$, *and* $\sigma > 0$. *If Y is another random variable such that* $y = e^x$, *derive the probability density function of Y.*

Sol. The functional relation is $y = e^x$ or $\ln y = x$. Also when $x = -\infty, y = 0$ and when $x = \infty, y = \infty$.

$$\frac{dx}{dy} = \frac{1}{y} \quad \text{and} \quad \left|\frac{dx}{dy}\right| = \frac{1}{y}.$$

$$g(y) = f(x)\left|\frac{dx}{dy}\right| = \frac{1}{y} \cdot \frac{1}{\sigma\sqrt{2\pi}} e^{-(x-\mu)^2/2\sigma^2}$$

Replacing x by $\ln y$ in the above equation, we get

$$g(y) = \frac{1}{y\,\sigma\sqrt{2\pi}} e^{-(\ln y - \mu)^2/2\sigma^2}\ ;\ \text{for } 0 \le y \le \infty.$$

In this example the distribution of X is normal and the distribution of Y is log-normal. More about normal and lognormal distributions can be found in later chapters.

EXERCISES

2.1. Construct the sample space for the experiment of observing the number of rainy weeks at a given place in a year. (If the total rainfall exceeds 1 mm in a week, then it may be considered a rainy week). Show the following events on the sample space. (*i*) A = {the number of rainy weeks is less than 20} (*ii*) B = {the number of rainy weeks is more than 15 but less than 25}. (*iii*) $C = A \cup B$ and (*iv*) $D = A \cap B$. How do you alter the sketching of your sample space to represent the event that the number of rainy weeks is even.

2.2. Consider the experiment of measuring the daily temperature and relative humidity of the atmosphere at a weather station. Suppose that the temperature on any day varies between 5°C and 50°C and the relative humidity between 0 and 100%. Sketch the sample space for this experiment. If the temperature is between 20°C and 30°C and at the same time humidity is between 55 and 65%, then it is defined to be comfortable weather. Indicate the event on the sample space that the day is of comfortable weather.

2.3. Find the union $A_1 \cup A_2$ and the intersection $A_1 \cap A_2$ of the two sets A_1 and A_2 for the following cases

(*i*) $A_1 = \{x : 0 < x < 2\}$ and $A_2 = \{x : 1 \le x < 3\}$

(*ii*) $A_1 = \{(x, y) : 0 < x < 2, 0 < y < 2\}$ and $A_2 = \{(x, y) : 1 < x < 3, 1 < y < 3\}$.

2.4. If A and B are two sets in the sample space S such that $A \subset B$, then show that $B^c \subset A^c$.

2.5. Show that $(A^c \cup B) \cap (A^c \cap B) = A^c$.

2.6. Show that $(A \cup B) \cap (A \cap B) = (A \cap B^c) \cup (B \cap A^c)$.

2.7. Show that, if $P[A] = P[B] = P[A \cap B]$, then

$$P[A \cap B^c) \cup (B \cap A^c)] = 0.$$

2.8. Show that, if $P[A] = P[B] = 1$, then $P[A \cap B] = 1$.

2.9. Which of the following relationships are true ?

(*i*) $(A \cup B) = (A \cap B^c) \cup B$ (*ii*) $A \cup B = (A^c \cap B)$

(*iii*) $(A \cup B) = (A \cap B) \cup B$ (*iv*) $(A \cup B)^c \cap C = A^c \cap B^c \cap C^c$.

2.10. A major river has two tributaries. The probabilities that it rains in either of the tributaries is 0.6. The probability that it rains in each tributary is 0.4. If a severe flood is caused by rainfall in both the tributaries, determine the probability that a severe flood occurs in the river. What is the probability that the rainfall occurs in one tributary given that the rainfall occurred in the other ?

2.11. Events A and B are mutually exclusive. Can they be independent ?

2.12. Using the information given in Example 2.8, find the probabilities of the following events, given the initial day is a dry day. (*i*) Next two days will be dry days (*ii*) Next two days will be rainy days. (*iii*) First day is rainy day and second day is a dry day. (*iv*) First day is dry day and the second day is rainy day.

2.13. Weather and river flow forecasting information is sent by routes R_1, R_2 and R_3 each with the same probability of being used. The probability of an error to be introduced in transmitting message on these routes are 1/5, 1/4 and 1/3 respectively. If a message is received without error, what is the probability that it was sent via. route 3.

2.14. A rainfall station has been in operation for 3 years. During this period there has been only one rainy day in November. On the basis of this data what can one say about the probability of having a rainy day at this station in November ?

2.15. The observed annual flood peaks of a stream for a period of 20 years in m^3/s are given below. Determine the probability that an annual peak in excess of 1500 m^3/s will occur in 2 successive years.

1340, 1380, 1450, 618, 523, 528, 1220, 1780, 1060, 412
184, 1480, 876, 113, 516, 1090, 944, 397, 282, 353.

2.16. Give some examples of discrete random variables and continuous random variables in hydrology.

2.17. If daily rainfall amounts at raingauge stations A and B are dependent variables, are rainy days at A and non-rainy days at B dependent or independent random events ?

2.18. An experiment consists of throwing two fair dies and the random variable X is defined as the sum of the two faces that show up. Sketch the probability mass function and the cumulative distribution function for this random variable.

2.19. Check whether the function given by $p(x) = \frac{x + 2}{25}$ for $x = 1, 2, 3, 4, 5$ can serve as the probability function of a discrete random variable ?

2.20. The age of a person at his death in years is a random variable X, whose probability density function is given by

$$f(x) = 3 \times 10^{-9} x^2 (1-x)^2 \text{ for } 0 \le x \le 100$$
$$= 0 \qquad \text{elsewhere}$$

What is the probability that a person dies between 60 and 70 years? What is the probability that he dies between 60 and 70 years given that he was alive at 60?

2.21. If the probability density function of a random variable X is given by

$$f(x) = k.\, e^{-3x} \text{ for } x \ge 0$$
$$= 0 \qquad \text{elsewhere.}$$

Find the value of k. Also obtain the cumulative distribution function.

2.22. The life span of a convective storm (measured in seconds) is a random variable whose probability density function is given by

$$f(x) = \frac{20000}{(x+100)^3} \qquad \text{for } x > 0$$
$$= 0 \qquad \text{elsewhere.}$$

Find the probability that the storm has a life span of (*i*) at least 210 seconds, (*ii*) at most 90 seconds and (*iii*) anywhere between 90 to 120 seconds.

2.23. The probability density function of daily rainfall may be assumed as given below.

prob. $[X = 0] = 0.25$

$$f(x) = \frac{x}{4} \qquad \text{for } 0 < x < 1.732$$

$$f(x) = 0.866 - \frac{x}{4} \qquad \text{for } 1.732 < x < 3.464$$

Is this a proper probability density function. If yes, what is $P[X > 0.5]$ and what is $P[X > 0.5/X \ne 0]$?

2.24. If two random variables X and Y have the joint probability density function given by

$$f(x, y) = \frac{2x + y}{4} \qquad \text{for } 0 < x < 1,\ 0 < y < 2$$
$$= 0 \qquad \text{elsewhere}$$

find the marginal density of X, marginal density of Y and also whether the two random variables are independent.

2.25. Given the joint probability density function

$$f(x, y) = \frac{2x + 8y}{5} \qquad \text{for } 0 < x < 1,\ 0 < y < 1$$
$$= 0 \qquad \text{elsewhere}$$

Find the probability that $(X + Y) \le 1.5$. What is the probability that $0 < X < 0.75$ given $Y = 0.5$?

2.26. A random variable X has the probability density function defined by

$$f(x) = x\,e^{-x^2/2} \quad 0 \le x < \infty$$
$$= 0 \quad \text{elsewhere}$$

and is related to another variable Y through $y = x^2$. Find the probability density function of Y.

2.27. The random variable X is of continuous type with cumulative distribution function $F(x)$. If Y is a random variable such that $Y = F(x)$, show that Y is uniformly distributed in the interval (0, 1).

2.28. The probability density function of a random variable X is given by

$$f(x) = a\,e^{-ax} \text{ for } 0 \le x < \infty$$
$$= 0 \quad \text{elsewhere.}$$

If $y = e^x$, find the probability density function of Y.

3

Some Characteristics of Distributions

3.1. INTRODUCTION

The complete knowledge of the probability distribution function of a random variable is needed, strictly speaking, to make predictions about its occurrence. But certain characteristics of the distribution like mean, variance etc., which summarise its dominant features are often very useful and may provide considerable information regarding the behaviour of the random variable. These characteristics are also sometimes referred to as the *descriptors* of the distribution. As will be seen in the following discussion mean and variance are related to the first and second moments of the distribution. The characteristics related to higher moments such as Skewness coefficient and Kurtosis coefficient will also be discussed in this chapter.

3.2. MEAN

The *mean* is one of the descriptors related to the central tendency, that is, the value around which the distribution is centered. The other descriptors of central tendency are the *median* and the *mode*. The mean is also known as the *arithmetic mean*, the *average*, or the *expected value*. It is denoted by μ or $E[X]$. If X is a discrete random variable with probability distribution function $p(x)$, then its mean is given by

$$\mu = E[X] = \sum_{\text{all } x_i} p(x_i) \cdot x_i \qquad \text{...(3.1)}$$

On the other hand, if X is a continuous random variable with probability density function $f(x)$, then its mean is given by

$$\mu = E[X] = \int_{-\infty}^{\infty} x \cdot f(x) \cdot dx \qquad \text{...(3.2)}$$

As can be seen from this definition, mean is nothing but the centroid of the probability density function. It may also be noted here that, in the case of discrete random variable, the expected value need not be an integer though the random variable itself can take on only integer values. For example, in the case of rolling a die, we get

$$E[X] = \sum_{i=1}^{6} x_i \, p(x_i) \text{ for } i = 1, 2, \ldots\ldots, 6.$$

Since $p(x_i) = 1/6$ for all x_i and $x_i = i$, $E[X] = (1 + 2 + 3 + 4 + 5 + 6]/6 = 3.5$, which is not an integer.

We can note that, if the vertical line $x = a$ is an axis of symmetry of $f(x)$, then $E[X] = a$ since the centroid has to lie on the axis of symmetry. In particular, if $f(-x) = f(x)$, then $E[X] = 0$ since the line of symmetry in this case will be $x = 0$.

The median in the case of a continuous random variable is defined as a value z such that

$$\int_{-\infty}^{z} f(x) \, dx = 0.5 \qquad \ldots(3.3)$$

Therefore the median is that value such that half the area under the probability density function is to its left and the remaining half to its right. The median for a discrete random variable is defined similarly except that the integral is replaced by the summation over the values of the random variable. The median is often an appropriate measure of central tendency especially for variables that are not symmetrically distributed. This is because it is not sensitive to the extreme values.

The mode in the case of discrete random variable is defined as that value which has the highest probability, whereas in the continuous type it is that value at which the probability density function has the maximum ordinate (ridge). Sometimes, a distribution may have more than one mode. A distribution with only one mode is known as *unimodal* distribution.

If the distribution is symmetric, then the mean and median coincide. If the distribution is symmetric and unimodal, then the mean, median and mode will coincide.

If Y is a function of a random variable X given by $Y = g(X)$, then Y is also a random variable. The expectation of Y for a discrete case is given by

$$E[Y] = E[g(X)] = \sum_{\text{for all } x_i} g(x_i) \, p(x_i) \qquad \ldots(3.4)$$

In the continuous case it is given by

$$E[Y] = E[g(X)] = \int_{-\infty}^{\infty} g(x) f(x)\, dx \qquad ...(3.5)$$

The following important results are the logical deductions of Eqs. (3.4) and (3.5) (*i*) the expected value of a constant C equals the constant itself and (*ii*) the expected value of a constant times the random variable equals the constant times the expected value of the random variable. That is

$$E[C] = C \qquad ...(3.6)$$

$$E[CX] = CE[X] \qquad ...(3.7)$$

3.3. VARIANCE

The *variance* of the distribution is denoted by σ^2 or Var (X) and is defined as

$$\text{Var}(X) = \sigma^2 = \sum_i (x_i - \mu)^2 p(x_i) \text{ for discrete case} \qquad ...(3.8)$$

$$\text{Var}(X) = \sigma^2 = \int_{-\infty}^{\infty} (x - \mu)^2 f(x)\, dx \text{ for continuous case} \qquad ...(3.9)$$

Variance is a measure of *dispersion* or *spread* of the distribution. It is nothing but the second moment (to be defined later) of the distribution about its mean. The positive square root of the variance is called the *standard deviation* and is denoted by σ. Thus

$$\sigma = \{\text{Var}(X)\}^{1/2} \qquad ...(3.10)$$

The advantage of using standard deviation is that it has the same units as the original random variable. The variance of a distribution may be interpreted as the moment of inertia of the probability density function about the mean. The following properties of variance are very useful. (*i*) The variance of a random variable is unaffected when a constant is added to or subtracted from it. If Y is defined as a new random variable such that $Y = X \pm C$, then Var (Y) = Var (X). (*ii*) If a random variable is multiplied by a constant C, then the variance of the product is equal to the product of C^2 and the variance of the random variable. That is, if $Y = CX$, then

$$\text{Var}(Y) = C^2 \text{Var}(X).$$

If X is a random variable with mean μ and variance σ^2, and if we define a new random variable Z such that $Z = (X - \mu)/\sigma$, then it can be easily verified that Z will have mean of 0 and variance of 1. Then Z is called the *standardised variable.* Quite often it may be convenient to deal with the standardised variable instead of the original variable. The standardised variable Z has the probability density function which is exactly of the same shape as that of probability density function of X. In other words, if X is normally distributed, Z is also normally distributed, and if X follows gamma distribution, Z also follows gamma distribution. (Different types of distributions are discussed in later chapters). The standardisation can be made for any hydrologic variable, though there is sometimes incorrect understanding that primarily only normally distributed variables are standardised. This may be because the variable Z is often called a *normalised variable*. The disadvantage with standardisation is that originally only positive valued variable may become negative after standardisation which precludes taking logarithms.

To facilitate comparison among a number of random variables of different units, a characteristic called *coefficient of variation* is used. It is defined as the ratio of standard deviation to mean and is denoted by η_v.

$$\eta_v = \frac{\sigma}{\mu} \qquad \text{...(3.11)}$$

Eq. (3.11) is applicable only when both μ and σ exist. The coefficient of variation is extensively used in hydrology. Since the coefficient of variation is influenced by the shift in origin, it is not a convenient parameter for comparison of distributions when the origin is arbitrarily selected without any physical meaning.

3.4. CHEBYSHEV'S INEQUALITY

Though the exact probability density function of a random variable is unknown, we can still workout the approximate probabilities using Chebyshev's (also known as Tchebycheff's) inequality provided that the mean and standard deviation of the variable are known. The Chebyshev's inequality states that, for any $k > 1$

$$P\,[(\mu - k\sigma) \le X \le (\mu + k\sigma)] \ge \left(1 - \frac{1}{k^2}\right) \qquad \text{...(3.12)}$$

In words, the inequality states that the probability that a random variable X shall fall into an internal whose end points are $(\mu \pm k\sigma)$ must at least be equal to $\left(1 - \frac{1}{k^2}\right)$ regardless of the underlying distribution of the variable. The above inequality can be easily verified as follows.

$$\sigma^2 = \text{Var}\,(X) = \int_{-\infty}^{\infty} (x - \mu)^2 f(x)\, dx$$

$$\sigma^2 = \int_{-\infty}^{\mu - k\sigma} (x - \mu)^2 f(x)\, dx + \int_{\mu - k\sigma}^{\mu + k\sigma} (x - \mu)^2 f(x)\, dx + \int_{\mu + k\sigma}^{\infty} (x - \mu)^2 f(x)\, dx$$

Since the integral $(x - \mu)^2 f(x)\, dx$ is non-negative, we can form the inequality

$$\sigma^2 \geq \int_{-\infty}^{\mu - k\sigma} (x - \mu)^2 f(x)\, dx + \int_{\mu + k\sigma}^{\infty} (x - \mu)^2 f(x)\, dx$$

by deleting the second integral. Now in the range ($-\infty$ to $\mu - k\sigma$) and also in the range ($\mu + k\sigma$ to ∞) $|\, x - \mu \,|$ is always greater that $k\sigma$. Hence we can write

$$\sigma^2 \geq k^2\sigma^2 \int_{-\infty}^{\mu - k\sigma} f(x)\, dx + k^2\sigma^2 \int_{\mu + k\sigma}^{\infty} f(x)\, dx$$

$$\geq k^2 \sigma^2 \{P[X \leq (\mu - k\sigma)] + P[X \geq (\mu + k\sigma)]\}$$

or $$\frac{1}{k^2} \geq P[\,|\, X - \mu \,|\, \geq k\sigma]$$

which is equivalent to

$$P[\,|\, X - \mu \,|\, \leq k\sigma] \geq 1 - \frac{1}{k^2}$$

which proves the inequality. The fact that rough probability statements can be made with so little information emphasises the importance of the standard deviation (or variance) as the measure of dispersion.

Example 3.1. *The annual runoff of a stream has a mean and standard deviation of 5000 and 1000 hectare-metres respectively. Find the ranges within which the runoff will lie with probability of at least 0.5, 0.75 and 0.90.*

Sol. From the Chebyshev's inequality, we have

$$P[(\mu - k\sigma) \leq X \leq (\mu + k\sigma)] \geq 1 - \frac{1}{k^2}$$

(*i*) When this probability has to be at least 0.5, the value of k^2 has to be 2, or $k = \sqrt{2}$. Therefore

$$P[\mu - \sigma\sqrt{2}) \leq X \leq (\mu + \sigma\sqrt{2})] \geq 0.5$$

$$P[(5000 - 1000\sqrt{2}) \leq X \leq (5000 + 1000\sqrt{2})] \geq 0.5$$

$$P[3585.8 \leq X \leq 6414.2] \geq 0.5$$

This means that annual runoff in any year will be between 3585.8 and 6414.2 hectare-metres with a minimum probability of 0.5.

(*ii*) Similarly for a minimum probability of 0.75, we get

$$P[3000 \leq X \leq 7000] \geq 0.75$$

(*iii*) and for a minimum probability of 0.9, we get

$$P[1837.7 \le X \le 8162.3] \ge 0.9$$

The above answers are very conservative because we are making the predictions without using any underlying distribution. In later chapters we again look at this problem making use of the actual probability distribution.

Example 3.2. *If the probability density function of a random variable is given by*

$$f(x) = 6x\,(1-x) \quad \text{for } 0 < x < 1$$
$$= 0 \quad \text{elsewhere}$$

find the probability that X will take on a value within 1.5 standard deviations of the mean and compare it with the lower bound provided by Chebyshev's inequality.

Sol. First let the mean and variance of the distribution be determined.

$$\mu = \int_{-\infty}^{\infty} x\,f(x)\,dx = \int_{0}^{1} x\,.\,6x\,(1-x)\,dx = 0.5$$

$$\sigma^2 = \int_{-\infty}^{\infty} (x-\mu)^2 f(x)\,dx = \int_{0}^{1} (x-0.5)^2\, 6x\,(1-x)\,dx = \frac{1}{20}$$

$$\sigma = \frac{1}{\sqrt{20}} = 0.2236.$$

Now, the probability that X will take on a value within 1.5 standard deviations of the mean is the probability that it will take on a value between $(0.5 - 0.2236 \times 1.5)$ and $(0.5 + 0.2236 \times 1.5)$, that is, between 0.1646 and 0.8354.

$$P[0.1646 \le X \le 0.8354] = \int_{0.1646}^{0.8354} 6x\,(1-x)\,dx = 0.855$$

Whereas the minimum probability (lower bound) from Chebyshev's inequality is obtained as

$$P[(\mu - 1.5\,\sigma) \le X \le (\mu + 1.5\,\sigma)] \ge 1 - \frac{1}{1.5^2}$$

$$\ge 0.5555$$

Observe that the statement "the probability is 0.855" is much stronger than "the probability is at least 0.5555".

3.5. COVARIANCE

In a manner analogous to the variance of a single random variable, the *covariance* between two random variables X and Y is defined by

$$\text{Cov}\,(X, Y) = E\,[(X - \mu_x)\,.\,(Y - \mu_y)] \qquad \text{...(3.13)}$$

or $$\text{Cov}(X, Y) = \int_{-\infty}^{\infty}\int_{-\infty}^{\infty} (x-\mu_x)(y-\mu_y) f(x,y)\, dx\, dy \quad \text{...(3.14)}$$

where μ_x and μ_y are the means of the variables X and Y which are obtained from their respective marginal distributions and $f(x, y)$ is the joint probability density function of the random variables X and Y. Eq. (3.13) may also be written as

$$\text{Cov}(X, Y) = E[XY] - E[X] \,.\, E[Y] \quad \text{...(3.15)}$$

If the random variables are independent, then we have $f(x, y) = f(x) \,.\, f(y)$. Thus, for the case of independent random variables, Eq. (3.14) can be written as

$$\text{Cov}(X, Y) = \int_{-\infty}^{\infty}\int_{-\infty}^{\infty} (x-\mu_x)(y-\mu_y) f(x) f(y)\, dx\, dy$$

$$= \int_{-\infty}^{\infty} (x-\mu_x) f(x)\, dx \,.\, \int_{-\infty}^{\infty} (y-\mu_y) f(y)\, dy$$

$$= (\mu_x - \mu_x)(\mu_y - \mu_y) = 0$$

Therefore when the random variables are independent their covariance is zero. Making use of this condition in Eq. (3.15) one can say that when the random variables are independent

$$E[XY] = E[X] \,.\, E[Y] \quad \text{...(3.16)}$$

The *coefficient of correlation* denoted by ρ between two random variables X and Y is defined by

$$\rho = \frac{\text{Cov}(X, Y)}{\{\text{Var}(X) \,.\, \text{Var}(Y)\}^{1/2}} = \frac{\text{Cov}(X, Y)}{\sigma_x \,.\, \sigma_y} \quad \text{...(3.17)}$$

It may be noted that the coefficient of correlation is unitless and hence dimensionless. It is a measure of *linear dependence* between the variables X and Y. It can be shown that the value of ρ for any two random variables always lies between -1 and $+1$, that is, $-1 \le \rho \le 1$. Consider the expected value of the quantity $\{(X-\mu_x)/\sigma_x + t\,(Y-\mu_y)/\sigma_y\}^2$ where t is any real number. This expectation must be greater than zero since it is the expectation of a squared quantity which is always positive. Thus

$$E\left[\left\{\frac{(X-\mu_x)}{\sigma_x} + \frac{t\,(Y-\mu_y)}{\sigma_y}\right\}^2\right] \ge 0$$

$$E\left[\left(\frac{X-\mu_x}{\sigma_x}\right)^2\right] + 2t\,E\left[\left(\frac{X-\mu_x}{\sigma_x}\right)\cdot\left(\frac{Y-\mu_y}{\sigma_y}\right)\right]$$

$$+ t^2 E\left[\left(\frac{Y-\mu_y}{\sigma_y}\right)^2\right] \ge 0$$

$$1 + 2t\rho + t^2 \geq 0$$

An expression of quadratic form $ap^2 + bp + c > 0$ is true only when the condition $(b^2 - 4ac) < 0$ is satisfied. Applying this condition to the present situation, we obtain

$$(4\rho^2 - 4) < 0 \;;\; \text{or } (\rho^2 - 1) < 0 \;;\; \text{or } \rho^2 < 1$$

$$-1 \leq \rho \leq 1 \qquad ...(3.18)$$

The correlation coefficient ρ measures how the random variables X and Y vary together. When the random variables are independent then $\rho = 0$ since Cov (X, Y) is zero. However, if $\rho = 0$ it only implies that there is no linear dependence between the variables, but there may be other type of dependence. When $\rho = -1$, the variables are said to be *perfectly negatively correlated* and when $\rho = 1$, the variables are said to be *perfectly positively correlated*. We shall discuss correlation coefficient in detail again in Chapter 7.

3.6. MOMENTS

Apart from mean and variance, the other descriptors like the *Skewness coefficient* and the *Kurtosis coefficient* are defined in terms of the moments of the distribution. Hence, we now introduce the concept of moment. The rth moment of the distribution about the origin, denoted by μ'_r, is defined as the expectation of X^r. Therefore,

$$\mu_r' = E[X^r] = \sum_{\text{all } x_i} x_i^r \, p(x_i) \text{ for } r = 0, 1, 2, \ldots\ldots \qquad ...(3.19)$$

when X is a discrete random variable, and

$$\mu_r' = E[X^r] = \int_{-\infty}^{\infty} x^r f(x)\, dx \quad \text{for } r = 0, 1, 2, \ldots\ldots \qquad ...(3.20)$$

when X is a continuous random variable. It is obvious that the first moment about the origin is nothing but the mean of the distribution. That is, $\mu_1' = \mu$. The moments about the mean are more useful than the moments about the origin. The rth moment about the mean, often simply called the rth *central moment* is denoted by μ_r and is defined as the expectation of $(X - \mu)^r$. Therefore, in the case of discrete random variable

$$\mu_r = E[(X - \mu)^r] = \sum_{\text{all } x_i} (x_i - \mu)^r \, p(x_i) \qquad ...(3.21)$$

and in the case of continuous random variable it is given by

$$\mu_r = E\,[(X - \mu)^r] = \int_{-\infty}^{\infty} (x - \mu)^r f(x)\, dx \qquad ...(3.22)$$

A distribution is completely specified when all its moments are known. However, many distributions can be adequately described by the first 4 moments. In addition to providing useful descriptive measures, the first 4 moments play an important role in fitting empirical distributions and approximating the distribution of a random variable. These techniques will be discussed in Chapter 6. It can be easily verified that the first central moment is zero.

$$\mu_1 = E[(X - \mu)] = E[X] - E[\mu]$$

$$\mu_1 = \mu - \mu = 0 \qquad ...(3.23)$$

The second central moment is nothing but the variance σ^2.

$$\mu_2 = E[(X - \mu)^2] = \sigma^2 \qquad ...(3.24)$$

The second central moment can also be expressed in terms of the second moment about the origin.

$$\mu_2 = E[(X - \mu)^2] = E[(X^2 - 2\mu X + \mu^2)]$$

$$= E[X]^2 - 2\mu\, E[X] + E[\mu^2]$$

$$= \mu_2' - 2\mu^2 + \mu^2$$

$$\therefore \quad \mu_2 = \mu_2' - \mu^2 \qquad ...(3.25)$$

The significance of σ^2 as a measure of dispersion has already been discussed.

Skewness coefficient. The third central moment is related to the *assymetry* or *skewness* of a distribution. To render the measure of assymetry independent of the dimension of the variable, the *skewness coefficient* of the distribution is defined as the ratio of the third central moment to the second central moment raised to the power 1.5 and is denoted by γ_1. Thus,

$$\gamma_1 = \frac{\mu_3}{(\mu_2)^{1.5}} = \frac{\mu_3}{(\sigma^2)^{1.5}} = \frac{\mu_3}{\sigma^3} \qquad ..(3.26)$$

Sometimes it may be convenient to evaluate μ_3 in terms of the moments about the origin using the following relation.

$$\mu_3 = \mu_3' - 3\,\mu_2'\,\mu_1' + 2(\mu_1')^3 \qquad ...(3.27)$$

Pearson's skewness coefficient is one of the other measures of the assymetry which is frequently used. It is given by

$$\gamma_1 = \frac{\mu - m}{\sigma} \qquad ...(3.28)$$

in which m is the mode of the distribution. If mode can be accurately determined, this is a convenient definition for γ_1.

A single peaked distribution with a long tail on the left hand side will have skewness coefficient less than zero ($\gamma_1 < 0$). Such a distribution is said to be *negatively skewed* (or skewed to the right) distribution. And the distribution with a long tail on the right hand side will have skewness coefficient more than zero ($\gamma_1 > 0$) and it is said to be *positively skewed* (or skewed to the left). For symmetrical distributions the skewness coefficient is zero ($\gamma_1 = 0$). This is because

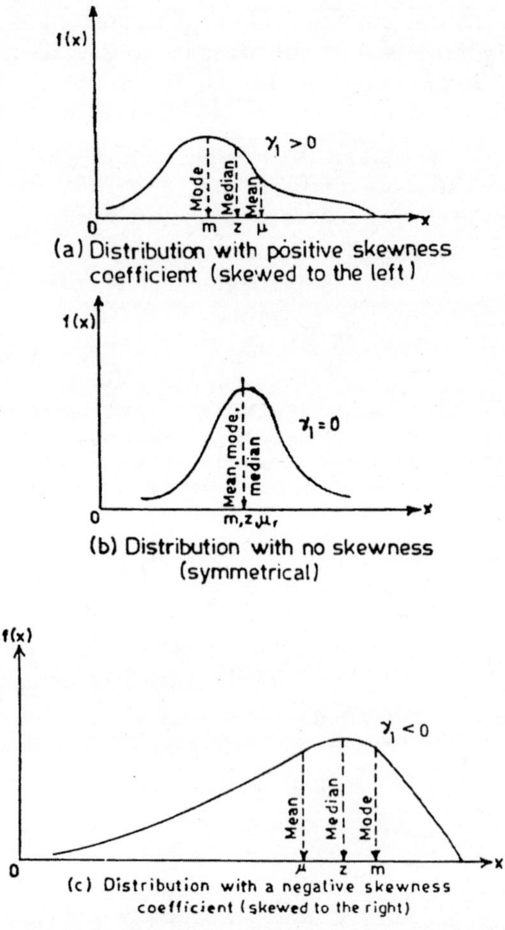

Fig. 3.1. Relation between mean, mode, median and skewness.

in the case of symmetric distributions μ_3 (in fact any odd central moment) will be zero. The distributions which are positively skewed, symmetrical and negatively skewed are shown in Fig. 3.1.

Kurtosis Coefficient. The fourth central moment is related to the peakedness of the distribution. The *kurtosis coefficient*, taken to be a measure of peakedness, is defined as the ratio of the fourth central moment to the square of the second central moment. It is denoted by γ_2. Thus

$$\gamma_2 = \frac{\mu_4}{(\mu_2)^2} = \frac{\mu_4}{(\sigma^2)^2} = \frac{\mu_4}{\sigma^4} \qquad ...(3.29)$$

The fourth central moment may be evaluated in terms of the moments about the origin using the relation

$$\mu_4 = \mu_4' - 4\,\mu_3'\,\mu_1' + 6\,\mu_2'\,(\mu_1')^2 - 3\,(\mu_1')^4 \quad ...(3.30)$$

For the normal distribution (to be discussed in Chapter 4) the kurtosis coefficient is 3. The normal distribution is said to be *mesokurtic.* If a distribution has a relatively greater concentration of probability near the mean than in the normal distribution, the kurtosis coefficient will be greater than 3 and the distribution is said to be *leptokurtic.* If a distribution has a relatively smaller concentration of probability near the mean than in the normal distribution, the kurtosis coefficient will be less than 3 and the distribution is said to be *platykurtic.* The *coefficient of excess*, denoted by E, of a given distribution is defined as the difference between its own kurtosis coefficient and the kurtosis coefficient of the normal distribution. That is

$$E = \gamma_2 - 3 \quad ...(3.31)$$

Therefore for a normal distribution $E = 0$, for a leptokurtic distribution $E > 0$, and for a platykurtic distribution $E < 0$.

It may be noted that both skewness coefficient and the kurtosis coefficient are dimensionless numbers. They are independent of the units of the original random variable.

Example 3.3. *A random variable is exponentially distributed if its probability density function is given by*

$$f(x) = \lambda e^{-\lambda x} \quad \text{for } x \geq 0$$
$$= 0 \quad \text{elsewhere.}$$

where λ is the parameter of the distribution. Find the mean, median, mode, standard deviation, skewness coefficient, kurtosis coefficient and the coefficient of excess for this distribution.

Sol. $$\mu = \int_{-\infty}^{\infty} x\, f(x)\, dx = \int_0^{\infty} x\, \lambda e^{-\lambda x}\,.\, dx = \frac{1}{\lambda}$$

If z is the median of the distribution, then

$$\int_0^z \lambda e^{-\lambda x},\, dx = 0.5 \text{ which yields } z = \frac{\ln 2}{\lambda}$$

where ln stands for the logarithm to the naperian base. The mode of the distribution is zero, since the probability density function has maximum ordinate at $x = 0$.

$$\mu_2 = \int_0^{\infty} \left(x - \frac{1}{\lambda}\right)^2 \lambda e^{-\lambda x}\,.\, dx = \frac{1}{\lambda^2}$$

$$\mu_3 = \int_0^{\infty} \left(x - \frac{1}{\lambda}\right)^3 \lambda e^{-\lambda x}\, dx = \frac{2}{\lambda^3}$$

$$\mu_4 = \int_0^{\infty} \left(x - \frac{1}{\lambda} \right)^4 \lambda e^{-\lambda x}\, dx = \frac{9}{\lambda^4}$$

Now,

$$\sigma = \sqrt{\mu_2} = \sqrt{1/\lambda^2} = \frac{1}{\lambda}$$

$$\gamma_1 = \frac{\mu_3}{\sigma^3} = \frac{(2/\lambda^3)}{(1/\lambda^3)} = 2$$

$$\gamma_2 = \frac{\mu_4}{\sigma^4} = \frac{(9/\lambda^4)}{(1/\lambda^4)} = 9$$

$$E = \gamma_2 - 3 = 9 - 3 = 6.$$

Hence for this distribution, we have

mean $\mu = 1/\lambda$; median $z = \ln 2/\lambda$

mode $m = 0$; standard deviation $\sigma = 1/\lambda$

skewness coefficient $\gamma_1 = 2$; kurtosis coefficient $\gamma_2 = 9$

excess coefficient $E = 6$.

3.7. MOMENT-GENERATING FUNCTION

Though the moments of most distributions can be determined directly by evaluating the necessary summations or integrals, there is an alternative method which sometimes provides considerable simplifications. This method utilises the *moment-generating function*. If X is a discrete random variable, its moment-generating function, when it exists, is given by

$$M_X(t) = E[e^{tx}] = \sum_x e^{tx}\, p(x) \qquad \text{...(3.32)}$$

When X is a continuous random variable, then

$$M_X(t) = E[e^{tx}] = \int_{-\infty}^{\infty} e^{tx} f(x)\, dx \qquad \text{....(3.33)}$$

The rth moment about the origin μ_r' is then found to be the rth derivative of $M_X(t)$ with respect to t and evaluated at $t = 0$. That means

$$\mu_r' = \left. \frac{d^r M_X(t)}{dt^r} \right|_{t=0} \qquad \text{..(3.34)}$$

The proof of Eq. (3.34) is based on expanding e^{xt} into Maclaurin's series. For the discrete case, we therefore get

$$M_X(t) = \sum_x \left[1 + tx + \frac{t^2 x^2}{2\,!} + \frac{t^2 x^3}{3\,!} + \ldots\ldots \right] p(x)$$

$$= \Sigma\, p(x) + t\, \Sigma\, x\, p(x) + \frac{t^2}{2\,!} \Sigma x^2\, p(x) + \;....\; + \frac{t^r}{r\,!} \Sigma x^r\, p(x) + ...$$

$$= 1 + t\mu_1' + \frac{t^2}{2\,!}\mu_2' + \;......\; + \frac{t^r}{r\,!}\mu_r' + \;......$$

$$\frac{d^r M_X(t)}{dt^r} = \mu_r' + \text{terms of } t$$

$$\therefore \quad \left. \frac{d^r M_X(t)}{dt^r} \right|_{t=0} = \mu_r', \text{ which completes the proof.}$$

Example 3.4. *The probability distribution function p(x) of a discrete random variable is given by*

$$p(x) = \tfrac{1}{8}\,(3C_x) \quad \textit{for } x = 0, 1, 2, 3$$

Find the moment generating function of this variable and use it to determine μ_1' *and* μ_2' *and also its variance.*

Sol. $$M_X(t) = \sum_x e^{tx}\, p(x)$$

$$= e^{t.0}\frac{3C_0}{8} + e^{t.1}\frac{3C_1}{8} + e^{t.2}\frac{3C_2}{8} + e^{t.3}\frac{3C_3}{8}$$

$$= \tfrac{1}{8}\,[1 + 3e^t + 3e^{2t} + e^{3t}] = \tfrac{1}{8}\,(1 + e^t)^3.$$

$$\frac{dM_X(t)}{dt} = \tfrac{3}{8}\,(1 + e^t)^2\, e^t$$

$$\mu_1' = \left. \frac{dM_X(t)}{dt} \right|_{t=0} = \tfrac{3}{8}\,(1 + 1)^2 \,.\, 1 = \tfrac{3}{2}$$

$$\frac{d^2 M_X(t)}{dt^2} = \tfrac{3}{4}\,(1 + e^t)\, e^t \,.\, e^t + \tfrac{3}{8}\,(1 + e^t)^2\, e^t$$

$$= \tfrac{3}{4}\,(1 + e^t)\, e^{2t} + \tfrac{3}{8}\,(1 + e^t)^2\, e^t$$

$$\mu_2' = \left. \frac{d^2 M_X(t)}{dt^2} \right|_{t=0} = \tfrac{3}{4}\,(1 + 1) \,.\, 1 + \tfrac{3}{8}\,(1 + 1)^2 \,.\, 1$$

$$= \frac{3}{2} + \frac{3}{2} = 3$$

$$\sigma^2 = \mu_2 = \mu_2' - \mu^2 = \mu_2' - (\mu_1')^2$$

$$= 3 - \left(\frac{3}{2}\right)^2 = \frac{3}{4}.$$

EXERCISES

3.1. What is the least value of k in Chebyshev's inequality for which the probability that a random variable takes on a value between $(\mu - k\sigma)$ and $(\mu + k\,\sigma)$ is (*a*) at least 0.95 and (*b*) at least 0.99.

3.2. If the probability density function of a random variable X is given by $f(x) = 630\, x^4 (1-x)^4$ for $0 < x < 1$; $f(x) = 0$ elsewhere, find the probability that X will take on a value within 2 standard deviations of the mean and compare it with the lower bound given by Chebyshev's inequality.

3.3. A random variable has a mean of 12 and variance of 9 and an unknown probability distribution. Find the least probability with which it lies between 6 and 18.

3.4. The annual precipitation depth over a basin is a random variable denoted by X. Likewise let annual runoff depth from the same basin be a random variable denoted by Y. If $E[X] = 796$ mm, Var [X] = 190096 mm^2, and $Y = 1.26\,X - 666$, find $E[Y]$ and Var $[Y]$.

3.5. Show that $E[(X-a)^2] = \sigma_x^2 + (\mu_x' - a)^2$, where μ_x and σ_x^2 are the mean and variance of X respectively and a is any arbitrary constant.

3.6. If X, Y and Z are three independent random variables with expectations μ_x, μ_y, μ_z respectively show that $E[XYZ] = \mu_x \cdot \mu_y \cdot \mu_z$.

3.7. X and Y are two random variables with means μ_x and μ_y and variances σ_x^2 and σ_y^2. The standardised variables are $R = (X - \mu_x)/\sigma_x$ and $S = (Y - \mu_y)/\sigma_y$. Show that the covariance, Cov (R, S), between R and S is same as the correlation coefficient ρ between X and Y.

3.8. A random variable X has mean μ_x and variance σ_x^2. What are the mean and variance of the random variables Y and Z defined as $Y = X/\mu_x$ and $Z = X/\sigma_x$.

3.9. The annual rainfall at a place is found to have a mean 760 mm and standard deviation 450 mm. What is the range within which the annual rainfall would lie at least with a probability of 0.5 and 0.75 ?

3.10. The probability density function of a random variable is given by

$$f(x) = \begin{cases} kx & \text{for } 0 \le x \le 4 \\ k(8-x) & \text{for } 4 \le x \le 8 \\ 0 & \text{elsewhere.} \end{cases}$$

Evaluate the constant k. Also find the mean, median, mode, standard deviation, skewness coefficient, kurtosis coefficient and excess coefficient for this distribution.

3.11. Assume that the random variable X has the following distribution.

$$f(x) = \frac{a^b}{\Gamma b} x^{b-1} e^{-ax}, \quad 0 \le x < \infty,\ a > 0,\ b > 0\ ;$$
$$= 0 \qquad \text{elsewhere.}$$

Find the moment generating function for this distribution and hence determine μ_1' and μ_2'.

4

Continuous Distributions

4.1. INTRODUCTION

There are many functions which fulfil the conditions of being a probability distribution function. Few of such distribution functions which are commonly used in modelling hydrologic variables are now discussed. First, the distributions of a continuous random variable are presented in this chapter followed by the distributions of discrete random variables in Chapter 5.

4.2. NORMAL DISTRIBUTION

The normal distribution, also known as the *Gaussian distribution*, is one of the most frequently used distributions to model the random phenomenon. Its probability density function is given by

$$f(x) = \frac{1}{\sigma\sqrt{2\pi}} e^{-(x-\mu)^2/2\sigma^2} \qquad ...(4.1)$$

with $-\infty < x < \infty$, $-\infty < \mu < \infty$, $\sigma > 0$ and where μ and σ are called the parameters of the distribution. Specifically μ is called the *location parameter* and σ is called the *scale parameter*. Since $f(x)$ in Eq. (4.1) is symmetrical about the line $x = \mu$, we have that $E[X] = \mu$. Further, since the area of $f(x)$ equals 1, we have

$$\int_{-\infty}^{\infty} e^{-(x-\mu)^2/2\sigma^2} \, . \, dx = \sigma\sqrt{2\pi} \qquad ...(4.2)$$

Differentiating both sides of Eq. (4.2) with respect to σ, we obtain

$$\int_{-\infty}^{\infty} \frac{(x-\mu)^2}{\sigma^3} e^{-(x-\mu)^2/2\sigma^2} \, . \, dx = \sqrt{2\pi}$$

Multiplying both sides of the above equation by $\sigma^2/\sqrt{2\pi}$, we get

$$\int_{-\infty}^{\infty} (x-\mu)^2 \, . \, \frac{1}{\sigma\sqrt{2\pi}} e^{-(x-\mu)^2/2\sigma^2} \, . \, dx = \sigma^2.$$

The left hand side of the above equation is nothing but $E[(X-\mu)^2]$; that is Var (X).

Therefore we notice that the parameters μ and σ are nothing but the mean and standard deviation of the distribution. Since the normal distribution is symmetrical about the mean μ, all odd central moments are zero. Hence the skewness coefficient of the normal distribution is zero. By evaluating the fourth central moment, one can show that the kurtosis coefficient for the normal distribution is 3. Fig. 4.1 (a) shows two normal distributions with same standard

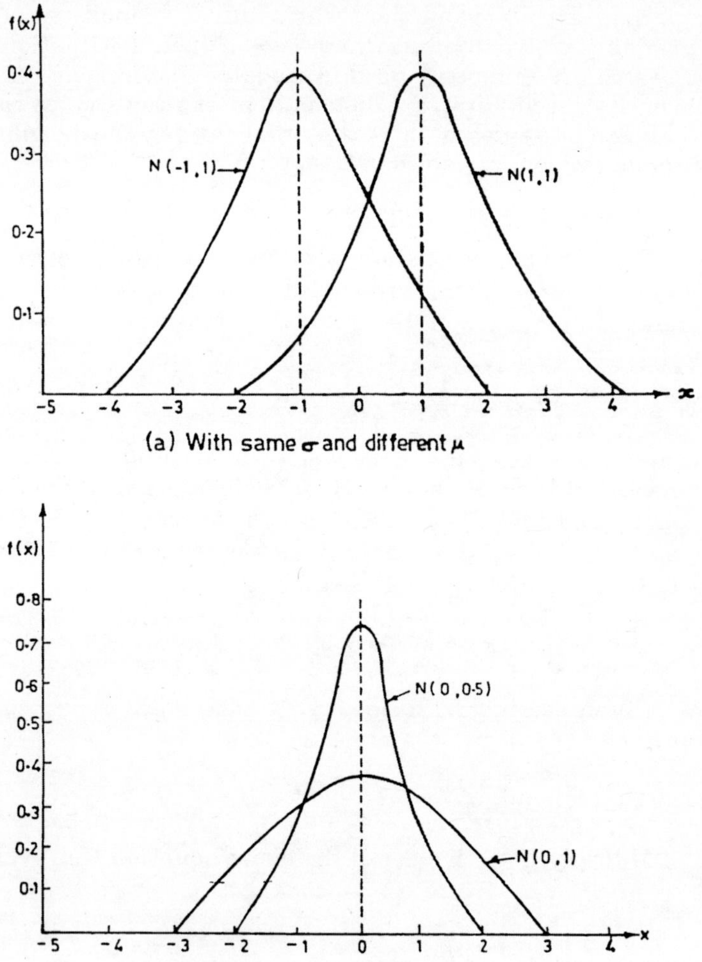

(a) With same σ and different μ

(b) With same μ and different σ

Fig. 4.1. Normal distribution.

deviation but different means and Fig. 4.1 (*b*) shows two normal distributions with same mean and different standard deviations. These figures explain why μ and σ are called the location and scale parameters respectively.

If a random variable X is normally distributed with mean μ and variance σ^2, then it is written that X is $N(\mu, \sigma^2)$. The probability that X takes on a value equal to or less than x can be obtained as

$$P[X \le x] = F(x) = \int_{-\infty}^{x} \frac{1}{\sigma\sqrt{2\pi}} e^{-(u-\mu)^2/2\sigma^2} \, . \, du \qquad ...(4.3)$$

Unfortunately Eq. (4.3) cannot be evaluated analytically. One has to resort to numerical methods and if the tabulation of the integral were to be made, a separate table would be required for each combination of μ and σ. If we use the transformation that $Z = (X - \mu)/\sigma$, then Z will have mean of 0 and variance of 1. The variable Z also follows normal distribution. That is, Z is $N(0, 1)$ and is said to be standardised normal. The probability density function of Z is given by

$$f(z) = \frac{1}{\sqrt{2\pi}} e^{-z^2/2} \qquad ...(4.4)$$

The cumulative distribution of Z is given by

$$P[Z \le z] = F(z) = \int_{-\infty}^{z} \frac{1}{\sqrt{2\pi}} e^{-u^2/2} \, . \, du \qquad ...(4.5)$$

Eq. (4.5) may be very conveniently used to determine the probabilities of any normal random variable with any mean and any standard deviation. Because

$$P[x_1 \le X \le x_2] = P[(x_1 - \mu)/\sigma \le Z \le (x_2 - \mu)/\sigma]$$
$$= P[z_1 \le Z \le z_2]$$
$$P[x_1 \le X \le x_2] = F(z_2) - F(z_1) \qquad ...(4.6)$$

$F(z)$ is widely tabulated*. As $f(z)$ is symmetrical, it is sufficient if $F(z)$ is tabulated for one half and it is usually tabulated for the right half as in Table A.1 in the Appendix. To deal with values of z on the left half, that is, for $z < 0$ we may use the relationship that $F(-z) = 1 - F(z)$. A study of Table A.1 reveals that 68.26% of the area under the normal curve is within 1 standard deviation, 95.44% within 2 standard deviations and 99.74% within 3 standard deviations from the mean. These are called 1, 2 and 3 sigma bounds of the normal distribution. The probability that a normal random variable takes on a value less than $(\mu - 3\sigma)$ is only $(1 - 0.9974)/2 = 0.0013$.

*For large positive value of z, $F(z) \approx 1 - f(z)/z$.

Therefore for any normal random variable with μ more than 3σ, the probability that it takes a negative value is negligible. This is the reason why the normal distribution is used to model a random variable with a lower bound of 0, though strictly speaking the lower bound of a normal random variable is theoretically $-\infty$.

The normal distribution has some very useful reproductive properties. If a random variable X is $N(\mu, \sigma^2)$ and $Y = a + bX$, then the distribution of Y can be shown to be $N(a + b\mu, b^2\sigma^2)$. Further if X_i, for $i = 1, 2, \ldots\ldots, n$ are independently and normally distributed with means μ_i and variances $\sigma_i^{\,2}$, then $Y = a + b_1X_1 + b_2X_2 + \ldots\ldots + b_nX_n$ is normally distributed with mean μ_y and variance $\sigma_y^{\,2}$, where $\mu_y = a + b_1\mu_1 + b_2\mu_2 + \ldots\ldots + b_n\mu_n$ and $\sigma_y^{\,2} = b_1^{\,2}\,\sigma_1^{\,2} + b_2^{\,2}\,\sigma_2^{\,2} + \ldots\ldots + b_n^{\,2}\,\sigma_n^{\,2}$. Any linear function of independent normal random variables is also a normal random variable.

Approximations to F(z). Though Table A.1 gives the accurate values of $F(z)$ for various values of z, the following approximations may be used to evaluate $F(z)$ for any z eliminating the need for interpolation.

$$B = \tfrac{1}{2}\{1 + 0.196854\,|\,z\,| + 0.115194\,|\,z\,|^2 + 0.000344\,|\,z\,|^3 + 0.019527\,|\,z\,|^4\}^{-4} \quad \ldots(4.7)$$

where $|\,z\,|$ is the absolute value of z. And now $F(z)$ is given by

$$\left.\begin{aligned} F(z) &= B && \text{for } z < 0 \\ &= 1 - B && \text{for } z \geq 0 \end{aligned}\right\} \quad \ldots(4.8)$$

The error in $F(z)$ obtained from Eqs (4.7) and (4.8) will be less than 0.00025.

Similarly the following approximation may be used to evaluate z for a given value of $F(z)$. Suppose $F(z)$ is specified. Then compute $p = 1 - F(z)$. In otherwords, p denotes the *exceedence probability* whereas $F(z)$ denotes the *cumulative* probability. An intermediate value w is defined by

$$w = \left[\ln\left(\frac{1}{p^2}\right)\right]^{1/2} \quad \text{for } 0 < p \leq 0.5 \quad \ldots(4.9)$$

Then z is calculated from

$$z = w - \frac{2.515517 + 0.802853\,w + 0.010328\,w^2}{1 + 1.432788\,w + 0.189269\,w^2 + 0.001308\,w^3} \quad \ldots(4.10)$$

If $p > 0.5$, then $(1-p)$ is substituted for p in Eq. (4.9) and the z obtained from Eq. (4.10) is given the negative sign. The error in this formula is less than 0.00045 in z.

These approximations will be very convenient especially for computer processing since the *table look-up* and the interpolations will be avoided.

Central limit theorem. Let $X_1, X_2, \ldots\ldots, X_n$ be n independent random variables with means $\mu_1, \mu_2, \ldots\ldots, \mu_n$ and variances $\sigma_1^2, \sigma_2^2, \ldots\ldots, \sigma_n^2$ whose individual distributions are not specified and may be different. Then the *central limit theorem* states that, under some general conditions a new random variable X defined as $X = X_1 + X_2 + \ldots\ldots + X_n$ will tend to follow a normal distribution as n approaches ∞ with mean $\mu = \mu_1 + \mu_2 + \ldots\ldots + \mu_n$ and variance $\sigma^2 = \sigma_1^2 + \sigma_2^2 + \ldots\ldots + \sigma_n^2$. The "general conditions" mentioned in the theorem are informally summarised as follows : Each X_i taken individually contribute a negligible amount to the variance of the sum and it is not likely that a single term makes a large contribution to the sum. (*i.e.*, there cannot be one or two dominating X_i in the sum). The immense practical importance of the central limit theorem lies in the fact that the statement of the theorem can be made without the exact knowledge of the individual distributions of contributing variables. It has been observed that, even if the variables in the summation are not independent but if they are large in number the summation still tends to follow normal distribution. If the variables are independent the convergence to normal would be faster.

The normal distribution is very useful in hydrology. Hydrologic random variables related to large time interval such as annual runoff, annual precipitation and similar variables are approximately normally distributed. If a daily event such as daily runoff or daily precipitation is considered to be a random variable, the corresponding annual event is a result of the summation of 365 random variables. Though the daily events are not truly independent, since 365 is a fairly large number, by invoking central limit theorem annual hydrologic variables can be taken to follow normal distribution. It is observed from experience that the normal distribution is a good model for annual runoff at many river gauging sites and for annual precipitation at many precipitation gauging stations all over the world. Random errors associated with any hydrologic measurements are usually symmetrically distributed and can be well approximated by the normal distribution. Normal distribution also plays an important role in testing the hypotheses since many of the sample statistics follow normal distribution.

In general, a random variable which is the outcome of a number of causative factors which are additive in effect can be approximated by a normal distribution on the basis of central limit theorem.

Example 4.1. *The annual runoff of a stream is modelled by a normal distribution with mean and standard deviation of 5000 and 1000 ha-m respectively. (i) Find the probability that the annual runoff in any year is more than 6500 ha.m. (ii) Find the probability that it would be between 3800 and 5800 ha.m.*

Sol. Let X be the random variable denoting the annual runoff. Then Z given by $(X - 5000)/1000$ is $N(0, 1)$.

$$(i)\ P[X \geq 6500] = P\left[Z \geq \frac{6500 - 5000}{1000}\right] = P[Z \geq 1.5]$$

$$= 1.0 - P[Z \leq 1.5] = 1.0 - F(1.5)$$

From Table A-1, we read that when $z = 1.5$, $F(z) = 0.9332$

$\therefore\ P(X \geq 6500] = 1.0 - 0.9332 = 0.0668.$

$$(ii)\ P[3800 \leq X \leq 5800] = P\left[\frac{3800 - 5000}{1000} \leq Z \leq \frac{5800 - 5000}{1000}\right]$$

$$= P[-1.2 \leq Z \leq 0.8] = F(0.8) - F(-1.2)$$

$$= F(0.8) - \{1 - F(1.2)\}$$

From Table A-1, we get $F(0.8) = 0.7881$ and $F(1.2) = 0.8849$

$\therefore\ P[3800 \leq X \leq 5800] = 0.7881 - (1.0 - 0.8849) = 0.673.$

Example 4.2. *For the normal random variable of Example 4.1, find the range which is symmetrical about mean within which the runoff will lie with a probability of 0.5 and compare the result with that obtained in Example 3.1.*

Sol. Let x_1 and x_2 be the end points of the range within which X will lie with a probability of 0.5. Then if $z_1 = (x_1 - 5000)/1000$ and $z_2 = (x_2 - 5000)/1000$, we need to find z_1 and z_2 such that $P[z_1 \leq Z \leq z_2] = 0.5$. or $F(z_2) - F(z_1) = 0.5$. Since the range is symmetrical about mean, $F(z_2)$ must be 0.75 and $F(z_1)$ must be 0.25. From Table A.1, when $F(z_2) = 0.75$, we get $z_2 = 0.675$ and again from symmetry z_1 must be -0.675.

$$z_2 = 0.675 = (x_2 - 5000)/1000\ ; \quad \text{or} \quad x_2 = 5675$$

$$z_1 = -0.675 = (x_1 - 5000)/1000\ ; \quad \text{or} \quad x_1 = 4325.$$

That means, the annual runoff will lie in the range 4325 to 5675 ha.m with a probability of 0.5. This range is smaller than 3585.8 to 6414.2 ha.m which we obtained in Example 3.1 using Chebyshev's inequality for a lower bound probability of 0.5. It is to be emphasized here that the Chebyshev's inequality makes no assumption about the distribution and also it gives only the lower bound of the probability and not the exact probability.

Example 4.3. *A random variable follows normal distribution and has a coefficient of variation of 0.4. Find the probability that it takes on a value less than zero.*

Sol. It is given that X is $N(\mu, \sigma^2)$ and $\sigma/\mu = 0.4$ or $\mu/\sigma = 2.5$. Now

$$P[X \le x] = P[Z \le (x - \mu)/\sigma]$$

$$\begin{aligned} P[X \le 0] &= P[Z \le (-\mu/\sigma)] = P[Z \le -2.5] \\ &= F[(-2.5)] = 1 - F(2.5) \\ &= 1.0 - 0.9938 = 0.0062 \end{aligned}$$

4.3. LOG-NORMAL DISTRIBUTION

A random variable X is said to be log-normally distributed if its probability density function is given by

$$f(x) = \frac{1}{x\,\sigma_n\sqrt{2\pi}}\, e^{-(\ln x - \mu_n)^2/2\sigma_n^{\,2}} \qquad ...(4.11)$$

for $0 < x < \infty$, $0 < \mu_n < \infty$ and $\sigma_n > 0$ in which μ_n and σ_n are the parameters of the distribution. By taking appropriate moments the mean, variance, Skewness coefficient and Kurtosis coefficient of this distribution can be shown to be given by the following equations.

Mean $\quad \mu = e^{\mu_n + 0.5\sigma_n^{\,2}}$ ...(4.12)

Variance $\quad \sigma^2 = e^{2\mu_n + \sigma_n^{\,2}}(e^{\sigma_n^{\,2}} - 1) = \mu^2 (e^{\sigma_n^{\,2}} - 1)$ ...(4.13)

Standard deviation $\quad \sigma = \mu\,(e^{\sigma_n^{\,2}} - 1)^{1/2}$ ...(4.14)

Coefficient of variation $\quad \eta_v = \frac{\sigma}{\mu} = (e^{\sigma_n^{\,2}} - 1)^{1/2}$ ...(4.15)

Skewness coefficient $\quad \gamma_1 = \eta_v^{\,3} + 3\,\eta_v$ ...(4.16)

Kurtosis coefficient $\quad \gamma_2 = \eta_v^{\,8} + 6\,\eta_v^{\,6} + 15\,\eta_v^{\,4} + 16\eta_v^2 + 3$ ...(4.17)

The mode of the distribution occurs at $(e^{\mu_n - \sigma_n^{\,2}})$ and the median at e^{μ_n}. After some manipulation, the parameters of the distribution μ_n and σ_n can be expressed in terms of the mean and variance as given below.

$$\mu_n = \frac{1}{2}\ln\left(\frac{\mu^4}{\mu^2 + \sigma^2}\right) \qquad ...(4.18)$$

$$\sigma_n^{\,2} = \ln\left(\frac{\sigma^2 + \mu^2}{\mu^2}\right) \qquad ...(4.19)$$

The log-normal distribution function is sketched for different combinations of μ_n and σ_n in Fig. 4.2. It can be noticed from this figure that μ_n and σ_n are not the *location* and *scale* parameters as in the case of normal distribution but they are, instead, the *scale* and *shape* parameters respectively. The distribution has positive skewness coefficient always (from Eq. 4.16), which increases with increase in σ_n. It is skewed to the left with a long exponential tail on the right. It is more peaked than normal distribution since its Kurtosis coefficient is always more than 3 (from Eq. 4.17).

If the lowers bound of the distribution is not zero but some positive value β, then the distribution function is modified as given below.

$$f(x) = \frac{1}{(x - \beta)\, \sigma_n \sqrt{2\pi}}\, e^{-[\ln (x - \beta) - \mu_n]^2 / 2\sigma_n^2} \qquad \text{...(4.20)}$$

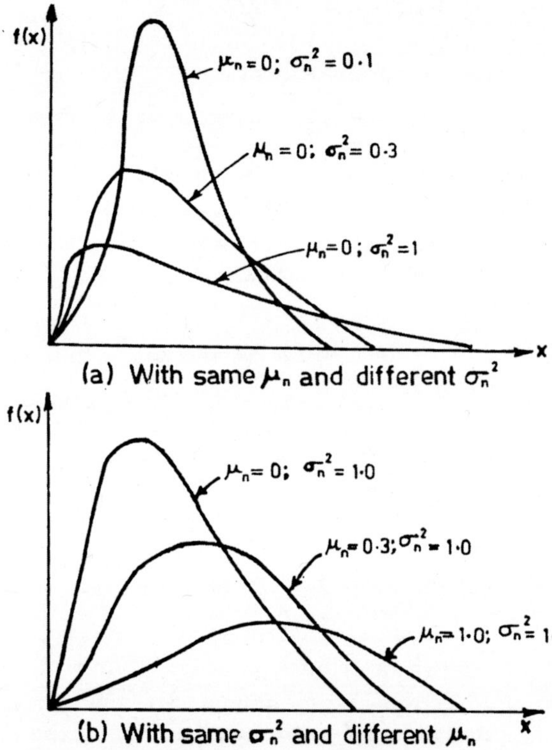

Fig. 4.2. Log-normal distribution.

It can be proved that the parameters μ_n and σ_n are the mean and variance of the variable ln X and also that ln X is normally distributed (see Example 2.14). In other words, if a variable is log normally distributed, then its logarithm is normally distributed. Because of this fact, no special tabulation of lognormal distribution is required. One can manage with the tabulation of normal distribution since

$$P[x_1 \le X \le x_2] = P\left[\frac{\ln x_1 - \mu_n}{\sigma_n} \le Z \le \frac{\ln x_2 - \mu_n}{\sigma_n}\right] \quad ..(4.21)$$

If X is equal to the product of many small effects, that is, if $X = X_1 . X_2 . X_3 .... X_n$, then $\ln X$ can be expected to be normally distributed. This can be seen by letting $Y = \ln X$ so that $Y = \ln (X_1 . X_2 .... X_n) = \ln X_1 + \ln X_2 + ... \ln X_n$. Since X_i are random variable $\ln X_i$ are also random variables and $Y = \ln X$ is a random variable made up of the sum of many other random variables. From the central limit theorem $\ln X$ can be expected to be normally distributed. Therefore the log-normal distribution can be used as the model for a process whose outcome is a result of number of causative factors having multiplicative effect. Many hydrologic variables such as precipitation, runoff, grain size of sediments and others are measured from the base zero and possess no upper bound. Since logarithm of zero is $-\infty$, the logarithmic transformation of these variables vary from $-\infty$ to $+\infty$. The application of the log-normal distribution to all such positive valued hydrologic variables is being practiced by many hydrologists. It has been found through experience that the lognormal distribution fits well to monthly runoffs, monthly precipitation and in some cases annual runoff and annual precipitation too. In the case of annual variables, though it is difficult to justify the choice of log-normal distribution, it will be accepted when it is found to be a better fit to the observed data than any other distribution.

Example 4.4. *The runoff of a stream in the month of October has a mean and standard deviation of 265 and 200 cumec-months respectively. Assuming that log-normal distribution is a good fit find the probability that October runoff in this stream in any year exceeds 350 cumec-months. What is the probability that the October runoff would fall in the range 150 to 350 cumec-months ?*

Sol. The parameters of the distribution μ_n and σ_n are evaluated from Eq. (4.18) and Eq. (4.19) using $\mu = 265$ and $\sigma = 200$ as follows.

$$\mu_n = \frac{1}{2}\ln\left(\frac{265^4}{265^2 + 200^2}\right) = 5.354$$

$$\sigma_n^2 = \ln\left(\frac{265^2 + 200^2}{265^2}\right) = 0.4508$$

$$\sigma_n = \sqrt{0.4508} = 0.671$$

Now $$P[X \ge 350] = P\left[Z \ge \frac{\ln 350 - 5.345}{0.671}\right] = P[Z \ge 0.751]$$

$$= 1 - F(0.751) = 1.0 - 0.7737$$
$$= 0.2263$$

and $P[150 \le X \le 350] = P\left[\frac{\ln 150 - 5.354}{0.671} \le Z \le \frac{\ln 350 - 5.354}{0.671}\right]$

$$= P[-0.512 \le Z \le 0.751]$$
$$= F(0.751) - F(-0.512)$$
$$= F(0.751) - \{1 - F(0.512)\}$$
$$= 0.7737 - (1 - 0.6957) = 0.4694.$$

Example 4.5. *A random variable has lognormal distribution with $\mu_n = 6.5$ and $\sigma_n = 0.6$. What are the mean and standard deviation of this variable. What is the probability that the random variable exceeds 450 ?*

Sol. Using Eq. (4.12) with $\mu_n = 6.5$ and $\sigma_n^2 = 0.36$

Mean $\qquad \mu = e^{(6.5 + 0.5 \times 0.36)} = e^{6.68} = 796.32$

and using Eq. (4.13)

Variance $\qquad \sigma^2 = \mu^2 (e^{\sigma_n^2} - 1) = (796.32)^2 (e^{0.36} - 1)$

$$= 274785.25$$

Standard deviation $\sigma = \sqrt{274785.25} = 524.2$

$$P[X \ge 450] = P\left[Z \ge \frac{\ln 450 - 6.5}{0.6}\right]$$
$$= P[Z \ge -0.651] = 1 - F(-0.651)$$
$$= 1 - \{1 - F(0.651)\}$$
$$= 1 - (1 - 0.7425) = 0.7425.$$

4.4. GAMMA DISTRIBUTION

A function used in the definition of gamma distribution, also known as *Erlang distribution*, is the gamma function. It is given by

$$\Gamma(n) = \int_0^{\infty} x^{n-1} e^{-x} dx \; ; \quad \text{for } n > 0 \qquad ...(4.22)$$

It can be shown that, when $n > 0$ $\lim\limits_{k \to \infty} \int_0^k x^{n-1} . e^{-x} dx$ exists.

By integrating Eq. (4.22) by parts it can be shown that

$$\Gamma(n) = (n - 1)\, \Gamma(n - 1) \qquad ...(4.23)$$

Also when $n = 1$, we have $\Gamma(1) = \int_0^{\infty} e^{-x} dx = 1$. Therefore, when n is an integer, Eq. (4.23) leads to the result that

$$\Gamma(n) = (n - 1)\,! \qquad ...(4.24)$$

Thus, the gamma function is a generalisation of the factorial. When n is not an integer, the gamma function has to be evaluated numerically. Usually, the gamma function is evaluated for all values only in the interval (1.0, 2.0) and for any value outside this interval the evaluation is accomplished using Eq. (4.23). Thus, for example, Γ (3.7) may be obtained from

$$\Gamma(3.7) = 2.7 \times 1.7 \times \Gamma(1.7)$$

where Γ (1.7) is obtained from the table of gamma function. Some salient values of gamma function are $\Gamma(0) = \infty$; $\Gamma(\frac{1}{2}) = \sqrt{\pi} = 1.7725$; $\Gamma(1) = \Gamma(2) = 1.0$.

A random variable X is said to follow gamma distribution if its probability density function is given by

$$f(x) = \frac{\lambda}{\Gamma(\eta)} \cdot (\lambda x)^{\eta - 1} e^{-\lambda x}; \quad \text{for } 0 < x \qquad ...(4.25)$$

where $\lambda > 0$ and $\eta > 0$ are called the parameters of the distribution. The gamma distribution function for various combinations of λ and η are shown in Fig. 4.3. As seen from this figure, gamma distribution function is quite flexible yielding a wide variety of shapes. In particular the distribution is a reverse J shaped curve for $\eta \leq 1$ and is single peaked for $\eta > 1$. Varying λ does not change the shape of

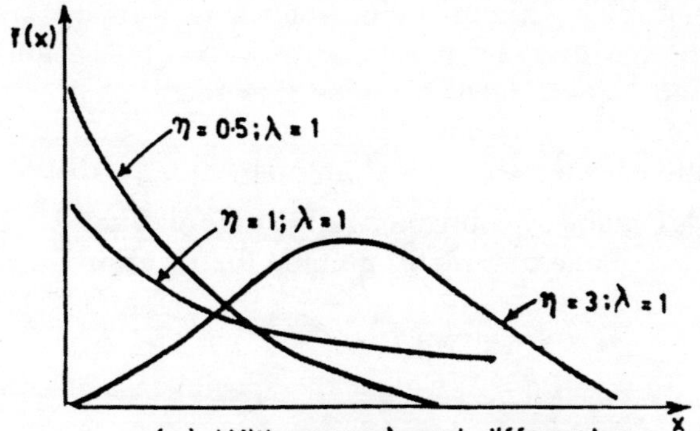

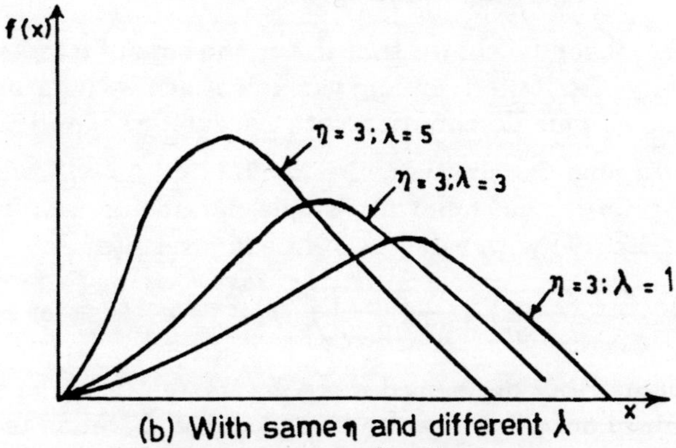

Fig. 4.3. Gamma distribution.

the distribution but only its scaling. For this reason, η and λ are called the *shape* and *scale* parameters respectively.

When the lower bound of the variable is not zero but a finite value β, then we get a three parameter gamma distribution given by

$$f(x) = \frac{\lambda}{\Gamma(\eta)} \left\{ \lambda (x - \beta) \right\}^{\eta - 1} e^{-\lambda (x - \beta)} \quad \text{for } x > \beta \qquad ...(4.26)$$

The descriptors of the gamma distribution are mean $\mu = \eta/\lambda$, variance $\sigma^2 = \eta/\lambda^2$, skewness coefficient $\gamma_1 = \frac{2}{\sqrt{\eta}}$ kurtosis coefficient $\gamma_2 = 3(\eta + 2)/\eta$. The parameters of the distribution η and λ can be written in terms of mean and variance as given below

$$\eta = (\mu/\sigma)^2 \qquad ...(4.27)$$

$$\lambda = \mu/\sigma^2 \qquad ...(4.28)$$

The use of two parameter and three parameter gamma distribution in hydrology is as common as the use of lognormal distribution. It has smoothly varying form and useful for describing skewed hydrologic variables without the need for logarithmic transformation. Though it cannot be justified theoretically, because of its flexibility many hydrologic variables like monthly precipitation, monthly runoff and even annual precipitation and annual runoff in some cases have been well approximated by gamma distribution.

Special cases of Gamma distribution. When $\lambda = 1$, we get the special case of one parameter gamma distribution.

$$f(x) = \frac{1}{\Gamma(\eta)} x^{\eta - 1} e^{-x} \qquad ...(4.29)$$

Likewise when $\eta = 1$, we get the special case of *exponential* distribution

$$f(x) = \lambda e^{-\lambda x} \qquad ...(4.30)$$

In fact it can be shown that if X is the sum of η independent exponentially distributed random variables each with parameter λ, then X has a gamma distribution with parameters η and λ.

For gamma distribution with $\lambda = 1/2$ and $\eta = \nu/2$, where ν is a positive integer, the following single-parameter family of *chi-square distribution* with ν degree of freedom results.

$$f(x) = \frac{1}{2^{\nu/2}\,\Gamma(\nu/2)} x^{\nu/2 - 1} e^{-x/2}; \quad \text{for } x > 0 \qquad ...(4.31)$$

This distribution is used more for the testing of hypotheses rather than to describe any random variable. It tends to normal distribution as v becomes large. This distribution function is tabulated in Table A.2 for various degrees of freedom. Its use will be explained while discussing the hypothesis testing of *goodness of fit* of any selected distribution to the observed data.

A very important property of gamma random variable, like that of a normal distribution, is that the sum of two gamma random variables also follows gamma distribution. This feature is of particular importance in generating synthetic hydrologic sequences.

4.5. BETA DISTRIBUTION

The probability density function of a random variable following *beta distribution* is given by

$$f(x) = \frac{x^{\alpha - 1} (1 - x)^{\beta - 1}}{B(\alpha, \beta)}; \quad \text{for } 0 < x < 1 \qquad \text{...(4.32)}$$

where $B(\alpha, \beta)$ is the beta function and $\alpha > 0$ and $\beta > 0$ are the parameters of the distribution. This distribution has both upper and lower bounds. The beta function is related to the gamma function as

$$B(\alpha, \beta) = \frac{\Gamma(\alpha)\, \Gamma(\beta)}{\Gamma(\alpha + \beta)} \qquad \text{...(4.33)}$$

The mean and variance of the distribution are given by

$$\mu = \frac{\alpha}{\alpha + \beta} \qquad \text{...(4.34)}$$

$$\sigma^2 = \frac{\alpha\beta}{(\alpha + \beta)^2 (\alpha + \beta + 1)} \qquad \text{...(4.35)}$$

Though the beta distribution is defined over the interval (0, 1), it can be transformed to any interval (a, b). The uniform distribution described next is a special case of the beta distribution with $\alpha = \beta = 1$.

4.6. UNIFORM DISTRIBUTION

A random variable is said to be uniformly distributed if the ordinate of its probability density function is a constant over the range of the variable. In order that the total area under probability density function is 1, if the range of the variable is from a to b, the constant value of the probability density function would be $1/(b - a)$ and therefore

$$f(x) = \frac{1}{(b - a)} \quad a \leq x \leq b \qquad \text{...(4.36)}$$

The uniform distribution is also called the *rectangular distribution* because of its shape. Its descriptors are mean $\mu = a + (b - a)/2$; variance $\sigma^2 = (b - a)^2/12$; skewness coefficient $\gamma_1 = 0$; kurtosis coefficient $\gamma_2 = 1.8$. The special case of uniform distribution arises when $a = 0$ and $b = 1$, which has a mean of 0.5 and variance of 1/12 while the skewness and kurtosis remain unaffected. Though the uniform distribution in its original form may not find any applications in hydrolcgy, it is essentially required to generate random numbers of any distribution to be used in stochastic simulation.

4.7. EXTREME VALUE DISTRIBUTION

Let $x_1, x_2, ..., x_n$ constitute a random sample of size n and let y be the extreme (largest or smallest) of the sample values. Then the probability distribution of this extreme value random variable will in general depend on the sample size and the parent distribution from which the sample was obtained. Three types of asymptotic (as $n \rightarrow$ large) distributions have been developed based on different parent distributions. These are

Type I : Parent distribution unbounded in the direction of the desired extreme and all moments of the distribution exist.

Type II : Parent distribution unbounded in the direction of the desired extreme and all moments of the distribution do not exist.

Type III : Parent distribution bounded in the direction of the desired extreme.

Interest may exist in the distribution of either the largest or the smallest extreme value. The type II extreme value distributions have found little application in hydrology. The type I distribution for the largest extreme volume is of importance to hydrologists since most hydrologic variables are unbounded on the right. Similarly type III distribution for the smallest extreme value is of interest to hydrologists since many hydrologic variables are bounded on the left by zero. These two extreme value distributions, that is type I for largest value and type III for smallest value, will now be discussed.

Extreme value type I distribution for largest value (Gumbel). The extreme value type I distribution for largest value is also known as the *Gumbel's extreme value distribution*, or the *Fisher-Tippett type I distribution*, or the *double exponential distribution*. The type I asymptotic distribution for maximum value is the limiting model as n approaches infinity when the n independent values have an initial distribution whose right tail is unbounded and which is an exponential type. The normal, log-normal, gamma and exponential distributions, all meet this requirement. The probability density function is given by

$$f(x) = \alpha \exp [-\alpha (x-\beta) - \exp \{-\alpha (x-\beta)] \quad ...(4.37)$$
$$-\infty < x < \infty ; -\infty < \beta < \infty, \alpha > 0.$$

The parameters α and β are the *scale* and *location* parameters with β being the mode of the distribution. The descriptors of the distribution are mean $\mu = (\beta + 0.5772/\alpha)$; variance $\sigma^2 = (\pi^2/6\ \alpha^2) = 1.645/\alpha^2$; skewness coefficient $\gamma_1 = 1.1396$; kurtosis coefficient $\gamma_2 = 5.4$. Like normal distribution, the skewness coefficient and kurtosis coefficient of this distribution are independent of the parameters of the distribution. The parameters α and β can be expressed in terms of the mean and variance as follows :

$$\alpha = 1.28255/\sigma \quad ..(4.38)$$
$$\beta = \mu - 0.45005\ \sigma \quad ...(4.39)$$

The cumulative distribution function is given by

$$F(x) = \exp [-\exp \{-\alpha (x-\beta))\}] \quad ...(4.40)$$

If the transformation

$$y = \alpha(x-\beta) \quad ...(4.41)$$

is used, the probability density function and the cumulative distribution function are expressed in the form

$$f(x) = \alpha \exp [-y - \exp (-y)] \quad ...(4.42)$$
$$F(x) = \exp [-\exp (-y)] \quad ...(4.43)$$

where y is called the *reduced variate*. The probability density function in terms of the reduced variate y is shown in Fig. 4.4.

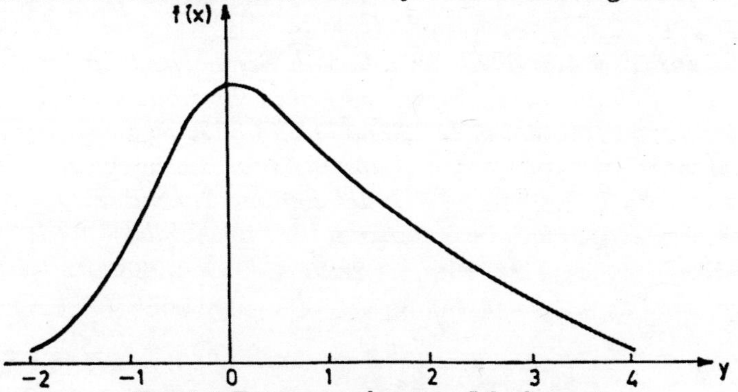

Fig. 4.4. Extreme value type I for largest value (Gumbel's) distribution.

The Gumbel's distribution is very extensively used to model the annual peak discharges all over the world. This is justified on the following reasons :

(*i*) The random variable daily discharge follows a distribution which has an exponential tail on the right.

(*ii*) The annual peak discharge is picked up as the largest of 365 daily discharges and the original number of elements in the sample, that is the 365 daily discharges, is sufficiently large for the asymptotic theory to be applicable.

(*iii*) The daily discharges are independent.

Although the last assumption clearly does not hold, it has been found from experience that extreme value distribution is a good fit for annual peak discharge.

Extreme value type III distribution for smallest value (Weibull). This distribution arises when the extreme is from a parent distribution that is limited in the direction of interest. It is also known as the *Weibull distribution.* It has found its greatest use in hydrology as the distribution of low stream flows (drought analysis) since the low flows are naturally bounded by zero on the left. The probability density function of this distribution is given by

$$f(x) = \frac{\alpha x^{\alpha-1}}{\beta^{\alpha}} \exp[-(x/\beta)^{\alpha}]\ x \geq 0, \alpha > 0, \beta > 0 \quad ...(4.44)$$

where α and β are the parameters of the distribution. The mean and variance are given by

$$\mu = \beta\, \Gamma(1 + 1/\alpha) \quad ...(4.45)$$

$$\sigma^2 = \beta^2 [\Gamma(1 + 2/\alpha) - \{\Gamma(1 + 1/\alpha)\}^2] \quad ...(4.46)$$

The cumulative distribution function is given by

$$F(x) = 1 - \exp[-(x/\beta)^{\alpha}] \quad ...(4.47)$$

Example 4.6. *The annual peak discharge of a stream is believed to follow Gumbel's extreme value distribution with mean and standard deviation of 10000 m^3/s and 3000 m^3/s respectively. (i) What is the probability that the annual peak discharge in any year is greater than 15000 m^3/s ? (ii) Also compute the magnitude of the annual peak which will be exceeded with a probability of 0.1.*

Sol. $\alpha = 1.28255/\sigma = 1.28255/3000 = 1/2339$

$\beta = \mu - 0.45005\sigma = 10000 - 0.45005 \times 3000 = 8650$

(*i*) $y = \alpha(x - \beta) = \frac{1}{2339}(15000 - 8650) = 2.7148$

$F(x) = \exp\{-\exp(-y)\} = \exp\{-\exp(-2.7148)\} = 0.9359.$

$P[X \geq 15000] = 1 - P[X \leq 15000] = 1 - F(x)$

$= 1 - 0.9359 = 0.0641$

(*ii*) $P[X \geq x] = 0.1 = 1 - F(x)$; or $F(x) = 0.9$

$0.9 = \exp\{-\exp(-y)\}$; $y = -\ln[-\ln(0.9)] = 2.25037$

$$y = \alpha (x + \beta) \text{ ; or } x = y/\alpha + \beta$$
$$x = 2.25037 \times 2339 + 8650 = 13913.7 \text{ m}^3/\text{s}.$$
$$P[X \geq 13913.7] = 0.1.$$

4.8. PEARSON DISTRIBUTIONS

The Pearson system of distributions are the solutions for $f(x)$ in an equation of the form

$$\frac{d\,[f(x)]}{dx} = \frac{f(x)\,(x - d)}{c_0 + c_1 x + c_2 x^2} \qquad ...(4.48)$$

where d is the mode of the distribution and c_0, c_1 and c_2 are the coefficients to be determined. It includes seven types of distributions. When $c_2 = 0$, the solution to Eq. (4.48) is a Pearson type III distribution. For $c_1 = c_2 = 0$, a normal distribution is the solution of Eq. (4.48). Thus the normal distribution is a special case of Pearson type III distribution. When ln X follows a Pearson type III distribution, then X is said to follow log-Pearson type III distribution.

Pearson type III distribution. The probability density function of this distribution is given by

$$f(x) = \frac{\lambda}{\Gamma(\eta)} \{ \lambda (x - \beta) \}^{\eta - 1} e^{-\lambda (x - \beta)}; \text{ for } x > \beta \qquad ...(4.49)$$

where λ, η and β are the parameters of the distribution and specifically β is the lower bound. This is nothing but the 3-parameter gamma distribution.

Log-Pearson type III distribution. This distribution is given by

$$f(x) = \frac{\lambda}{x\,\Gamma(\eta)} \{ \lambda (\ln x - \beta) \}^{\eta - 1} e^{-\lambda (\ln x - \beta)}; \text{ for } \ln x \geq \beta \qquad ...(4.50)$$

The Pearson type III distribution was applied in hydrology to describe the annual flood peaks when the data are very positively skewed (with high positive skewness coefficient). A log transformation is used to reduce the skewness and a log-Pearson type III distribution is fitted. The use of log-Pearson type III distribution is justified by the fact that it has been found to yield good results in many applications, particularly for flood peak data. This distribution is the standard distribution for the analysis of annual maximum flood in United States.

As a special case when ln X is symmetrical about its mean, log-Pearson type III distribution reduces to log-normal distribution.

4.9. OTHER CONTINUOUS DISTRIBUTIONS

There are several other continuous distributions like Rayleigh, Cauchy, Pareto, F, Student-t etc. Some of them are used only in the hypothesis testing and others do not find much applications in hydrology. A discussion on these distributions can be found in advanced text books on Probability and Statistics.

EXERCISES

4.1. Explain under what conditions does a random variable follow
(*i*) Normal distribution (*ii*) Log-normal distribution
(*iii*) Gumbel's distribution (*iv*) Weibull's distribution.

4.2. In the case of normal distribution, show that the two parameters μ and σ are the mean and variance of the distribution respectively.

4.3. How do you justify the selection of normal distribution to model the annual rainfall of a place ?

4.4. The annual peak runoff in a river has been modelled by a lognormal distribution with a lower bound = 15 m^3/s, $\mu_n = 5.03$ and $\sigma_n = 0.933$. Find the probability that annual runoff exceeds 400 m^3/s.

4.5. A random variable X is normally distributed with probability as given below :

$$P[X \leq 80] = 0.1 \text{ and } P[X \leq 320] = 0.9$$

Compute the mean and variance of X.

4.6. How do you justify the choice of Gumbel's extreme value distribution to describe the annual peak floods ?

4.7. The annual runoff of a stream has a mean and standard deviation of 256 m^3/s and 190 m^3/s. Find the probability that the annual runoff exceeds 400 m^3/s using (*i*) normal distribution (*ii*) Log-normal distribution, and Gumbel's distribution. Which answer do you think would be nearer to the correct.

4.8. Suggest suitable probability functions, with reasons, to model the following hydrologic variables. (*i*) annual runoff (*ii*) annual peak flood and (*iii*) monthly runoff.

4.9. The annual rainfall P over a basin is approximately normally distributed with mean 1150 mm and standard deviation 450 mm. Annual runoff R from the basin is related to annual rainfall by $R = 0.5\,P - 190$. What are the mean and standard deviation of annual runoff? What is the probability that the annual runoff will exceed 640 mm ?

4.10. Show that if X is $N(\mu, \sigma^2)$ then $y = a + b\,X$ is $N(a + b\mu, b^2\sigma^2)$.

4.11. Obtain the mean, variance, skewness and kurtosis coefficients of a uniformly distributed random variable in the range α to β.

4.12. Let X be a uniformly distributed and symmetrical about zero with variance equal to 1. Find the end values of the range of the variable α and β.

4.13. A random variable X is uniformly distributed on the interval (0, 2). Find the distribution of the random variable $Y = 5 + 2X$. What are the mean and variance of Y?

4.14. If the annual peak flood of a stream follows the Gumbel's distribution with mean 9000 m^3/s and standard deviation 4000 m^3/s, what is the probability that the annual peak flood exceeds 18000 m^3/s ? What is the probability that it will be utmost 15000 m^3/s ?

4.15. Repeat Exercise 4.14 assuming that annual peak flood follows log-normal distribution.

4.16. The annual peak flood of a stream exceeds 2400 m^3/s with a probability of 0.02 and exceeds 2730 m^3/s with a probability of 0.01. Assuming that it follows Gumbel distribution what is the probability that it exceeds 3000 m^3/s ? What is your answer if it is assumed that annual peak flood follows log-normal distribution ?

4.17. If peak discharge Q is log-normally distributed with mean μ_q and variance σ_q^2, what is the probability distribution of the peak stage S, when S and Q are related by $Q = aS^b$?

4.18. Repeat exercise 4.17 assuming that Q is distributed as type I extreme value distribution.

4.19. Is there an exponential probability density function that satisfies the condition $P[X \leq 2] = \frac{2}{3} P[X \leq 3]$. If so, find the value of λ.

[**Hint**. For exponential distribution $F(x) = 1 - e^{-\lambda x}$.]

4.20. If X_1 and X_2 follow gamma distribution with parameters η_1 and λ and η_2 and λ what is the distribution of $X = X_1 + X_2$.

5

Discrete Distributions

5.1. BINOMIAL DISTRIBUTION

Many problems involve independent repeated trials known as the *Bernoulli Trials* in which each trial results in one of the two possible outcomes frequently referred to as *success* and *failure* and where the probability of success, p, remains constant from one trial to the next. The situations in hydrology with such characteristics are rainy day and non-rainy day, or a flood greater than the design flood and a flood less than the design flood, etc., and similar such events. From the multiplicative law of independent events, the probability of a specified number of successes x in a given sequence of n trials [the other $(n - x)$ being failures] is equal to $p^x (1 - p)^{n-x}$ where p is the probability of success on a single trial, since the probability of failure on any single trial automatically becomes $(1 - p)$. From the rules of permutations and combinations there are $n!/[x!\,(n - x)!]$ equally likely sequences in which x successes and $(n - x)$ failures can occur in n trials. For simplicity this number is denoted by $\binom{n}{x}$. That is

$$\binom{n}{x} = \frac{n!}{x!\,(n-x)!} \qquad ...(5.1)$$

Consequently, from the law of probability of mutually exclusive events the probability of exactly x successes in n independent trials, with a probability of success p in a single trial, is given by the *binomial probability function*.

$$p(x) = P[X = x] = \binom{n}{x} p^x q^{n-x} \qquad ...(5.2)$$

where $0 \le p \le 1$, $q = 1 - p$ and $x = 0, 1, 2, ..., n$. That Eq. (5.2) represents a valid probability distribution can be easily verified as follows. Consider $(q + pt)^n$ which can be expanded using binomial coefficients as

$$(q + pt)^n = \binom{n}{0} q^n + \binom{n}{1} q^{n-1}(pt) + \binom{n}{2} q^{n-2}(pt)^2 + \ldots\ldots$$
$$+ \binom{n}{x} q^{n-x}(pt)^x + \ldots\ldots + \binom{n}{n}(pt)^n$$

$$(q + pt)^n = \sum_{x=0}^{n} \binom{n}{x} (pt)^x q^{n-x}$$

If we substitute $t = 1$ in the above equation, we get

$$(q + p)^n = 1 = \sum_{x=0}^{n} \binom{n}{x} p^x q^{n-x} \qquad \ldots(5.3)$$

This means that the sum of all the ordinates of the binomial distribution function given by Eq. (5.2) equals 1 and therefore it is a valid distribution function.

The mean of the distribution is evaluated as follows.

$$\mu = \sum_{x=0}^{n} x\, p(x) = \sum_{x=0}^{n} x \frac{n\,!}{x\,!\,(n-x)\,!} \cdot p^x q^{n-x}$$

$$= 0 + \sum_{x=1}^{n} x \frac{n\,!}{x\,!\,(n-x)\,!} p^x q^{n-x}$$

$$= \sum_{x=1}^{n} x \frac{n\,!}{x\,!\,(n-x)\,!} p^x q^{n-x}$$

$$= np \sum_{x=1}^{n} \frac{(n-1)\,!}{(x-1)\,!\,(n-x)\,!} p^{x-1} q^{n-x}$$

Now making the substitution that $y = (x - 1)$ and $m = n - 1$

$$\mu = np \sum_{y=0}^{m} \binom{m}{y} p^y q^{m-y}$$

Since the terms in the above summation represent the ordinates of the binomial distribution, the sum is equal to 1. Therefore, we have

$$\mu = np \qquad \ldots(5.4)$$

To find variance, first we determine μ_2' that is the second moment about the origin. This can be evaluated as

$$\mu_2' = E[X^2] = E[X(X-1)] + E[X].$$

We have already shown that $E[X] = \mu = np$. Now

$$E[X(X-1)] = \sum_{x=0}^{n} x(x-1) \binom{n}{x} p^x q^{n-x}$$

$$= 0 + \sum_{x=2}^{n} x(x-1) \frac{n!}{x!(n-x)!} p^x q^{n-x}$$

$$= \sum_{x=2}^{n} \frac{n!}{(x-2)!(n-x)!} p^x q^{n-x}$$

$$= n(n-1)p^2 \sum_{x=2}^{n} \frac{(n-2)!}{(x-2)!(n-x)!} p^{x-2} q^{n-x}.$$

Letting $y = x - 2$ and $m = n - 2$, we obtain

$$E[X(X-1)] = n(n-1)p^2 \sum_{y=0}^{m} \binom{m}{y} p^y q^{m-y}$$

$$= n(n-1)p^2 = n^2p^2 - np^2$$

$$\mu_2' = E[X(X-1)] + E[X] = n^2p^2 - np^2 + np$$

and finally, we have

$$\sigma^2 = \mu_2' - \mu^2 = n^2p^2 - np^2 + np - (np)^2$$

$$= np - np^2 = np(1-p) = npq$$

$$\sigma = \sqrt{npq} \qquad ...(5.5)$$

The skewness and kurtosis coefficients for binomial distribution can be shown to be

$$\gamma_1 = (q-p)/\sqrt{npq} \qquad ...(5.6)$$

$$\gamma_2 = 3 + (1 - 6pq)/npq \qquad ...(5.7)$$

The binomial distribution for different combinations of n and p are shown in Fig 5.1. It can be seen from this figure that when $p > 0.5$, the skewness coefficient is negative and the distribution is skewed to the right and when $p < 0.5$, the skewness coefficient is positive and the distribution is skewed to the left. When $p = 0.5$, the distribution is symmetrical and the skewness coefficient is zero. The distribution approaches symmetry as n becomes large even though $p \neq 0.5$.

The probability that there will be k or fewer successes in n independent trials is given by the cumulative distribution

$$F(k) = P[X \leq k] = \sum_{x=0}^{k} \binom{n}{x} p^x q^{n-x} \qquad ...(5.8)$$

Although the calculation of terms in the summation of Eq. (5.8) is straight forward it can be tedious especially when n and x are large and when there is a large number of terms in the summation. It can be shown that as n becomes large, the binomial distribution approaches normal distribution with same mean and variance. Therefore when n and x are large the normal distribution with $\mu = np$

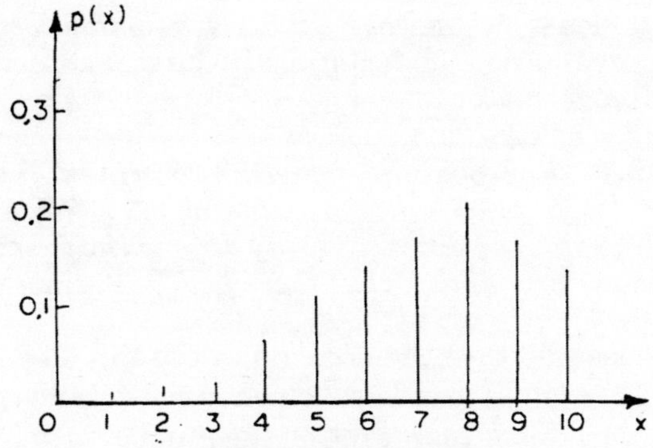

(a) With n : 10 and p : 0·75

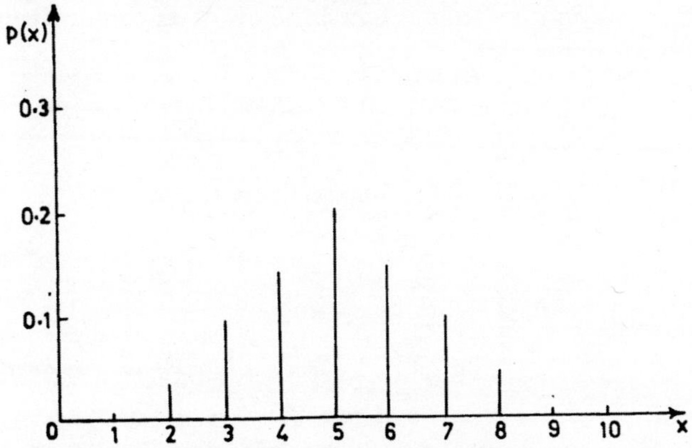

(b) With n=10 and p=0·50

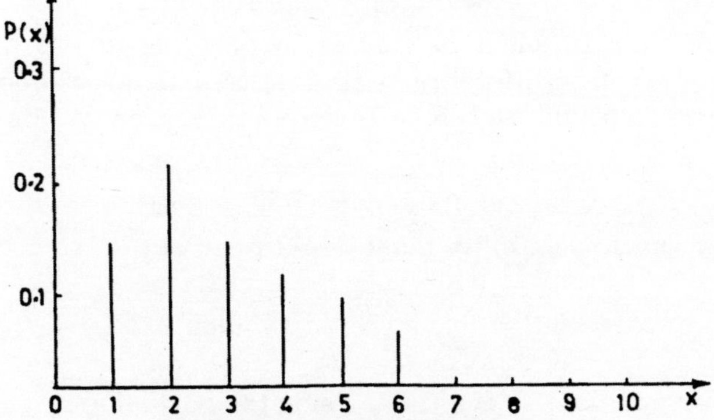

(c) With n=10 and p=0·25

Fig. 5.1. Binomial distribution.

and $\sigma = \sqrt{npq}$ may be conveniently used to find the required probability. This approximation gives reasonably accurate results if both

np and nq are at least 5. Another useful approximation which facilitates faster evaluation of factorials for large n especially for $n > 30$ is the *striling formula*

$$n\,! = n^n\, e^{-n} \sqrt{2\pi\, n} \qquad \text{...(5.9)}$$

Example 5.1. *The probability for a rainy day in the first week of October in any year is assigned a value of 0.35. What is the probability that there will be exactly 3 rainy days in the first week of October ? What is the probability that there will be at least 5 rainy days ?*

Sol. If we assume that the occurrence of rainy days in the first week of October is independent (which is not totally correct), then the random variable X representing the number of rainy days in that week is binomially distributed variable with $n = 7$ and $p = 0.35$ and the discrete values that it can take on in any year are 0, 1, 2, ..., 7. Therefore we have

$$P[X = x] = \binom{7}{x} 0.35^x\, 0.65^{7-x}.$$

(*i*) $$P[X = 3] = \binom{7}{3} 0.35^3\, 0.65^4 = 0.268$$

(*ii*) $$P[X \geq 5] = \sum_{x=5}^{7} \binom{7}{x} 0.35^x\, 0.65^{7-x}$$

$$= 0.0466 + 0.00836 + 0.0045 = 0.05946.$$

Example 5.2. *From the observations made over a long period of time the rainy day at a raingauge station is assigned a probability of 0.15 and the non-rainy day 0.85. What is the probability that there will be at least 60 rainy days in any year ?*

Sol. This problem is similar to part (*ii*) in the previous Example. The only difference is that since n is large (365 in the present case) we will use the normal approximation. We have

$$n = 365\ ;\ p = 0.15\ ;\ q = 0.85\ ;\ np = 365 \times 0.15 = 54.75$$

$$\sigma = \sqrt{npq} = \sqrt{365 \times 0.15 \times 0.85} = 6.82$$

The exact probability must be obtained as

$$P[X \geq 60] = \sum_{x=60}^{365} \binom{365}{x} 0.15^x\, 0.85^{365-x}$$

which is very tedious to evaluate. Using the normal approximation

$$P[X \geq 60] = P\left[Z \geq \frac{60 - 54.75}{6.82} \right] = P\,[Z \geq 0.77]$$

$$= 1 - F(0.77) = 1.0 - 0.7794 = 0.2206.$$

Example 5.3. *The magnitude of a T-year flood is that which is exceeded with a probability of (1 / T) in any year. A system is said to be designed for T-year flood (T is also called the return period or the recurrence interval) if it has a capacity which will be exceeded by a flood equal to or greater than T-year flood.*

For what return period a system has to be designed to provide 75 per cent assurance that failure would not occur in the next 20 years ?

Sol. Let T be the return period of the flood for which the system is designed. Then the probability that the annual peak flood exceeds the design flood in any year, p, is $(1/T)$ and the probability that the design flood is not exceeded in any year is $(1 - p)$ that is $(1 - 1/T)$. Let n be the life period of the project in years. The probability that the design flood is not exceeded continuously for the next n years is $(1 - p)^n$ or $(1 - 1/T)^n$.

The probability that the design flood is exceeded once or more in n years is $[1 - (1 - 1/T)^n]$. This is the risk involved in the system design. The assurance provided by designing the system for a T-year return period is (1 – risk), that is $(1 - 1/T))^n$, which is stipulated to be 0.75 in this problem.

$\therefore$ $0.75 = (1 - 1/T)^{20}$, which gives $T = 69.93 \approx 70$ years.

Hence, the system has to be designed for a 70 year return period.

5.2. NEGATIVE BINOMIAL DISTRIBUTION

This distribution is again related to the Bernoulli trials. We may be interested in the number of trial on which the kth success occurs. For instance we may be interested in the probability of the fifth occurrence of a 20 year flood on thirtieth year. If the kth success is to occur on the xth trial, there must be $(k - 1)$ successes in the first $(x - 1)$ trials and the probability for this is

$$\binom{x-1}{k-1} p^{k-1} q^{x-k}$$

and since the probability of a success on the kth trial is p, the probability that the kth success occurs on xth trial is given by

$$p \cdot \binom{x-1}{k-1} p^{k-1} q^{x-k} = \binom{x-1}{k-1} p^k q^{x-k}$$

A random variable X is said to have a *negative binomial distribution* (also known as *Pascal distribution*) if its probability distribution is given by

$$P[X = x] = p\ (x, k, p) = \binom{x-1}{k-1} p^k q^{x-k} \quad ...(5.10)$$

for $x = k, k + 1, k + 2, ...$

Thus, the number of the trial on which the kth success occurs is a random variable having a negative binomial distribution with parameters k and p. The name "negative binomial distribution" is due to the fact that the values of $p(x, k, p)$ for $x = k, k + 1, k + 2, ...$ are the successive terms of the binomial expansion of $\left(\frac{1}{p} - \frac{q}{p}\right)^{-k}$

The mean and the variance of the negative binomial distribution are given by

$$\mu = \frac{k}{p} \quad ...(5.11)$$

$$\sigma^2 = \frac{kq}{p^2} \quad ...(5.12)$$

Example 5.4. *What is the probability that the fifth occurrence of a 20 year flood will be on forty fifth year ?*

Sol. The year on which the fifth occurrence takes place is a negative binomial distribution.

$$P[X = x] = p(x, k, p) = \binom{x-1}{k-1} p^k q^{x-k}$$

Here $x = 45$; $k = 5$, and $p = \frac{1}{T} = \frac{1}{20} = 0.05$; so $q = 0.95$

$$P[X = 45] = \binom{44}{4} 0.05^5\ 0.95^{40} = 0.00545.$$

5.3. GEOMETRIC DISTRIBUTION

The geometric distribution also arises from a sequence of Bernoulli trials. Here the number of trials in the sequence is not specified, and the random variable X is defined to be the number of trials required till the first success is met with. The distribution is given by

$$p(x) = P[X = x] = q^{x-1}\ p. \quad ...(5.13)$$

It can be seen that the geometric distribution is a special case of negative binomial distribution with $k = 1$. It is easy to verify that this is a valid probability distribution since

$$\sum_{x=1}^{\infty} p\, q^{x-1} = p \sum_{k=0}^{\infty} q^k$$

$$= p\ (1 + q + q^2 + ...)$$

$$= p\ (1 - q)^{-1} \quad [\because\ q = 1 - p < 1]$$

$$= \frac{p}{(1-q)} = \frac{p}{p} = 1.$$

The mean and the variance of this distribution are

$$\mu = 1/p \qquad \text{...(5.14)}$$

$$\sigma^2 = q/p^2 \qquad \text{...(5.15)}$$

Example 5.5. *What is the probability that a 10 year flood will occur for the first time during the fifth year after the completion of a project ? What is the probability that it will be at least the fifth year before a 10 year flood occurs ?*

Sol. Let us say that the occurrence of flood is a success with a probability p. In the present case $p = 1/T = 1/10 = 0.1$ and therefore $q = 0.9$. If the flood occurs in the fifth year for the first time, then the time period before the first success is 4 years.

$$\therefore \qquad P[X = 5] = q^{5-1}\, p = (0.9)^4 \times 0.1 = 0.0656$$

The probability that it will be at least the fifth year before the first occurrence is equal to the probability of no occurrence in the first 4 years which is equal to $(0.9)^4 = 0.6561$.

Example 5.6. *What is the probability that exactly 9 years will elapse between occurrences of a 10 year flood.*

Sol. This is same as the probability of the first success in the tenth trial. That is, $x = 10$, $p = 1/T = 1/10 = 0.1$, and $q = 0.9$.

$$P[X = x] = q^{x-1}\, p$$

$$P[X = 10] = (0.9)^9 \times 0.1 \doteq 0.0387.$$

5.4. POISSON DISTRIBUTION

Poisson distribution is the limiting case of a binomial distribution which arises when n and p are independent and when n tends to be very large and p tends to be very small while their product np tends to be a finite value λ. That is,

$$\lim_{\substack{n \to \infty \\ p \to 0}} \binom{n}{x} p^x q^{n-x} = \frac{\lambda^x e^{-\lambda}}{x\,!}, \text{ where } np \to \lambda.$$

Therefore, for Poisson distribution, we have

$$P[X = x] = p(x) = \frac{\lambda^x e^{-\lambda}}{x\,!} \qquad \text{...(5.16)}$$

The descriptors of this distribution are

$$\mu = \lambda \qquad \text{...(5.17)}$$

$$\sigma^2 = \lambda \qquad \text{..(5.18)}$$

$$\gamma_1 = \frac{1}{\sqrt{\lambda}} \qquad \text{...(5.19)}$$

$$\gamma_2 = 3 + \frac{1}{\lambda} \qquad \text{...(5.20)}$$

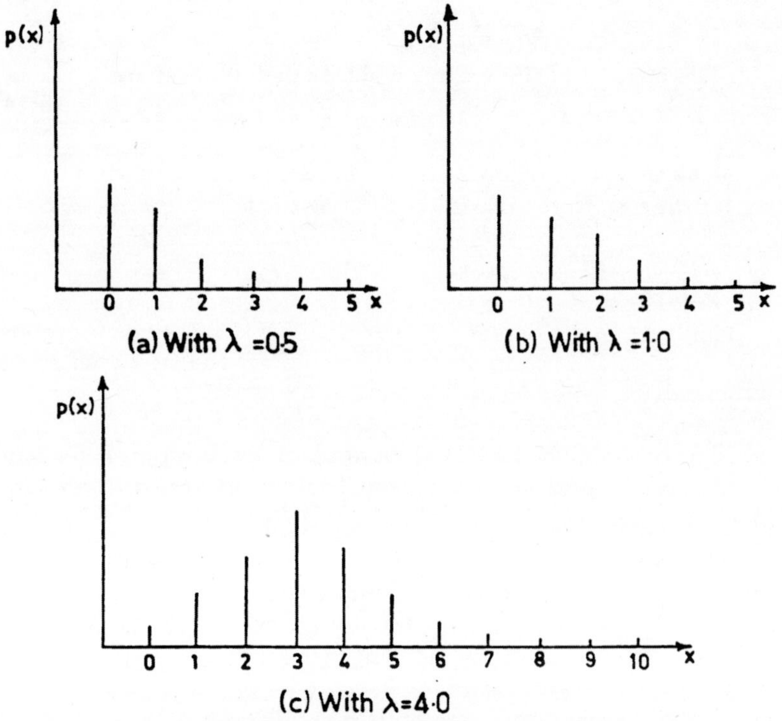

Fig. 5.2. Poisson distribution.

Fig. 5.2 shows the Poisson distribution for various value of λ. For $\lambda < 1$, the distribution is highly skewed to the left and for $\lambda > 1$, it is unimodal. It converges to normal as $\lambda \to \infty$.

Example 5.7. *A system is designed for a 25 year event. What is the probability that there will be fewer than 4 such events in the next 50 year period.*

Sol. Here $p = 1/T = 1/25 = 0.04$ and $n = 50$

Here n is relatively large and p is relatively small. This justifies the use of Poisson distribution in this example

$$\lambda = np = 50 \times 0.04 = 2$$

$$P[X < 4] = P[X \leq 3] = \sum_{x=0}^{3} \frac{2^x e^{-2}}{x\,!}$$

$$= e^{-2}\left(1 + 2 + 2 + \frac{8}{6}\right) = 0.857.$$

Example 5.8. *What is the probability that a flood with a return period of 25 years will occur twice in a 10 year period.*

Sol. We solve this example using both Poisson and binomial distributions. Here $n = 10$, $p = \frac{1}{T} = \frac{1}{25} = 0.04$, and $q = 0.96$.

If we use Poisson approximation, λ becomes $10 \times 0.04 = 0.40$

$$P[X = x] = \frac{\lambda^x e^{-\lambda}}{x\,!}$$

$$P[X = 2] = \frac{0.4^2 e^{-0.4}}{2\,!} = 0.0536$$

If we use the binomial distribution, the exact answer is obtained as

$$P[X = x] = \binom{n}{x} p^x q^{n-x}$$

$$P[X = 2] = \binom{10}{2} (0.04)^2 \times (0.96)^8 = 0.05194.$$

Thus we see that the solution provided by the poisson approximation, though not identical, for most practical purposes agrees with the solution by binomial distribution satisfactorily.

5.5. HYPERGEOMETRIC DISTRIBUTION

Let us consider a set of N elements of which k elements are having a particular attribute. In other words, *k elements belong to one group and $(N - k)$* elements belong to the second group. Now if we draw n of the N elements contained in the set without replacement, the probability that there will be x elements belonging to the first group in these n elements is given by the hypergeometric distribution.

$$P[X = x] = p\,(x, n, N, k) = \frac{\binom{k}{x}\binom{N-k}{n-x}}{\binom{N}{n}} \qquad \text{...(5.21)}$$

for $x = 0, 1, 2, \ldots\ldots, n$

$x \le k$ and $(n - x) \le (N - k)$

The mean and variance of the hypergeometric distribution are

$$\mu = \frac{nk}{N} \qquad \text{...(5.22)}$$

$$\sigma^2 = \frac{nk\,(N - k)\,(N - n)}{N^2\,(N - 1)} \qquad \text{...(5.23)}$$

Example 5.9. *There are 24 industries along a river which dump their effluent into it after treatment. If 6 of them are in the habit of not treating their effluent to the required standard, and if pollution control board authorities examine the effluent of 10 industries at random, what is the probability that 2 of the industries* *habitual of undertreating their effluent are included in this random sample.*

Sol. For this given situation, hypergeometric distribution can be applied. From the problem, we have

$$N = 24, k = 6, x = 2 \text{ and } n = 10$$

$$P[X = x] = \frac{\binom{k}{x}\binom{N-k}{n-x}}{\binom{N}{n}}.$$

$$P[X = 2] = \frac{\binom{6}{2}\binom{24-6}{10-2}}{\binom{24}{10}} = \frac{\binom{6}{2}\binom{18}{8}}{\binom{24}{10}}$$

$$= \frac{15 \times 43758}{1961256} = 0.3347.$$

Example 5.10. *In the month of June, 10 rainy days occurred. If a record of 12 days is selected at random and their climate condition analysed, what is the probability that 4 of these days will be rainy days.*

Sol. Hypergeometric distribution can be used for this situation with $N = 30$, $k = 10$, $n = 12$ and $x = 4$.

$$P[X = 4] = \frac{\binom{10}{4}\binom{30-10}{12-4}}{\binom{30}{12}} = \frac{\binom{10}{4}\binom{20}{8}}{\binom{30}{12}}$$

$$= \frac{210 \times 125970}{86493225} = 0.30585$$

5.6. OTHER DISCRETE DISTRIBUTIONS

Other discrete distributions include *modified Poisson distribution, truncated binomial distribution, multinomial distribution, negative hypergeometric distribution* etc., which may not find frequent applications in hydrology. For a detailed discussion on these distributions, advanced textbooks on Probability and Statistics may be consulted.

EXERCISES

5.1. Compute all the ordinates of a binomial distribution with (*i*) $n = 7$ and $p = 0.2$, and (*ii*) $n = 7$ and $p = 0.5$ and plot them.

5.2. A project is designed for a 50 year event. What is the assurance that it will not fail in the next 50 years ?

5.3. For what return period should a project be designed to provide 95% assurance that the project would not fail in the next 50 years ?

5.4. Keeping the assurance constant at 95%, find the return period of the event for which the project has to be designed if the life period of the project is 10, 20, 30, ..., 100 years and construct a curve relating the design return period and the life of the project. Construct similar curve for other levels of assurance such as 90%, 75% and 50%.

5.5. What are the features of a binomial distribution ? When is binomial distribution applicable in hydrology ? When is the normal approximation to binomial possible ?

5.6. If a project is designed for a 50 year flood, what is the probability that exactly 3 floods exceeding the design flood will occurs in a 50 year period. Assume that successive annual floods are independent.

5.7. The annual peak discharge of a stream follows Gumbel's distribution with mean 15000 m^3/s and standard deviation 4800 m^3/s. For what discharge would you design a structure on this stream to provide 80% assurance that failure would not occur within next 50 years ?

5.8. In the first week of November of any year, the probability for a rainy day is 0.75. What is the probability that there will be exactly 5 rainy days in that week ? What is the probability that there will be utmost 2 rainy days ?

5.9. What is the most probable value of X, when it is binomially distributed with $n = 4$ and $p = 0.5$.

5.10. Solve Exercise 5.8 using Poisson distribution. Are you satisfied with these answers ?

5.11. A dam is designed for a 100 year flood. What is the probability that a flood exceeding design flood occurs in 10th year after its completion ?

5.12. A project is designed for a 50 year flood. What is the probability that a flood exceeding design flood occurs for 4th time in 40th year ?

5.13. As part of an air pollution survey, an inspector decides to examine the exhaust of 6 trucks out of 24 trucks manufactured by a company. If 4 of the company's trucks emit excessive amounts of pollutants, what is the probability that none of them will be included in the inspector's sample ?

5.14. Derive the expressions for the mean and variance of the Poisson distribution.

5.15. Assuming the occurrence of a rainy day in a year is independent, the probability assigned for a rainy day is 0.165. Using normal approximation to the binomial distribution, find the probability that there will be at least 60 rainy days in a year and also the probability that there will be utmost 50 rainy days in a year.

6

Estimation and Probability Plotting

6.1. SAMPLE STATISTICS

So far, we have worked out the probabilities of a random variable assuming that the probability distribution function is known to us. In practice, however, we will not be knowing either the form of the probability function or its parameters except in cases where one can assume the probability function *a priori*. Therefore, we have a situation wherein the probability properties of the variables of interest are unknown but we wish to determine its probability function empirically. This task must be accomplished with the help of the data of the random variable observed in the past.

A collection of n independent observations on a continuous random variable X, which are denoted as $(x_1, x_2, \ldots\ldots, x_n)$ may be called a *sample*. In the case of a discrete random variable, let $x_1, x_2, \ldots\ldots, x_r$ be the r discrete values the random variable can take and let the number of times these values have been observed be $n_1, n_2, \ldots\ldots, n_r$ so that $n_1 + n_2 + \ldots\ldots + n_r = n$. Any value computed as a function of the sample data is called a *statistic* or more specifically a *sample statistic*. The mean, standard deviation etc. of a sample are defined in terms of *sample moments* in a manner analogous to the moments of the distribution. In contrast, the moments of the distribution are known as the *population moments*. In the case of a discrete random variable, the sample mean, denoted by $\bar{x}$, is given by

$$\bar{x} = \frac{1}{n} \sum_{i=1}^{r} n_i x_i \qquad \ldots(6.1)$$

whereas for a continuous case it is given by

$$\bar{x} = \frac{1}{n} \sum_{i=1}^{n} x_i \qquad \ldots(6.2)$$

The sample mean is taken to be an estimate of the population mean. That is $\hat{\mu}_x = \bar{x}$, where $\hat{\mu}_x$ is called an estimator of μ_x. The kth central moment of the sample for a discrete case is given by

$$m_k = \frac{1}{n} \sum_{i=1}^{r} n_i (x_i - \bar{x})^k \qquad ...(6.3)$$

and for a continuous case

$$m_k = \frac{1}{n} \sum_{i=1}^{n} (x_i - \bar{x})^k \qquad ...(6.4)$$

The sample variance $s_x{}^2$ is defined, for the discrete case, as

$$s_x{}^2 = \frac{1}{(n-1)} \sum_{i=1}^{r} n_i (x_i - \bar{x})^2 \qquad ...(6.5)$$

and for the continuous case, as

$$s_x{}^2 = \frac{1}{(n-1)} \sum_{i=1}^{n} (x_i - \bar{x})^2 \qquad ...(6.6)$$

It may be observed that in terms of the second central moment, the sample variance can be expressed as

$$s_x{}^2 = \frac{n}{n-1} \cdot m_2 \qquad ...(6.7)$$

When the sample size is large, the factor $n/(n-1)$ approaches unity and the sample variance will be equal to m_2 itself as in the case of population variance. The factor $n/(n-1)$ is introduced to make the sample standard deviation which is an estimator of the population variance, (that is $\hat{\sigma}_x{}^2 = s_x{}^2$) unbiased. The sample skewness coefficient denoted by g_1 is given by

$$g_1 = \frac{n^2}{(n-1)(n-2)} \frac{m_3}{(m_2)^{3/2}} \qquad ...(6.8)$$

and the sample kurtosis coefficient denoted by g_2 is given by

$$g_2 = \frac{n^3}{(n-1)(n-2)(n-3)} \frac{m_4}{(m_2)^2}; \qquad ...(6.9)$$

where g_1 and g_2 are taken to be the estimators of the corresponding population coefficients γ_1 and γ_2 respectively. That is, $\hat{\gamma}_1 = g_1$ and $\hat{\gamma}_2 = g_2$. As the sample size becomes larger and larger, that is as n tends to infinity, the sample moments tend to coincide with population moments. But when the sample size is small, the sample moments, especially the higher moments, are not reliable. Therefore too much emphasis cannot be given to g_1 and g_2 when they are computed from a small sample.

6.2. ESTIMATION OF PARAMETERS OF THE DISTRIBUTION

The general procedure for estimating a parameter is to obtain a random sample ($x_1, x_2, \ldots, x_n$) from the population of X and then use this random sample in the estimation of the parameter. In general, the probability density function of interest may contain certain parameters like $\alpha_1, \alpha_2, \ldots$ The function may be written as $f(x, \alpha_1, \alpha_2, \ldots)$. If parameters are estimated using sample, then $\hat{\alpha}_i$ an estimator for the parameter α_i is a function of the observations of the random variable, that is, $\hat{\alpha}_i$ is a function of random sample. Since $\hat{\alpha}_i$ is a function of random variables, $\hat{\alpha}_i$ itself is a random variable possessing a mean, variance and probability distribution. Regardless of the distribution of $\hat{\alpha}_i$ one can intuitively say that the larger the sample size the closes is $\hat{\alpha}_i$ to α_i and also if many samples were used for finding $\hat{\alpha}_i$, the average of these values should be equal to α_i. These are but two qualities of the estimator known as *consistency* and *unbiasedness*. The other qualities include the *efficiency* and the *sufficiency*.

An estimator $\hat{\alpha}$ of a parameter α is said to be unbiased if $E[\hat{\alpha}] = \alpha$. The bias if any is given by $E[\hat{\alpha}] - \alpha$.

An estimator $\hat{\alpha}$ of a parameter α is said to be consistent if the probability that $\hat{\alpha}$ differs from α by more than an arbitrary small constant ε approaches zero as the sample size approaches infinity.

An estimator $\hat{\alpha}$ is said to be most efficient estimator of the parameter α if it is unbiased and its variance is at least as small as that of any other unbiased estimator of α. For estimating α, the relative efficiency of one parameter $\hat{\alpha}_1$, with respect to another estimator $\hat{\alpha}_2$ is the ratio of Var $(\hat{\alpha}_2)$/Var $(\hat{\alpha}_1)$.

An estimator $\hat{\alpha}$ is said to be a sufficient estimator of the parameter α if $\hat{\alpha}$ uses all of the information relevant to α that is contained in the sample.

Detailed discussion on the above four properties of the estimators and the procedures for determining if an estimator has these properties can be found in advanced books on Mathematical Statistics.

There are many methods of estimating population parameters from the sample data. They include graphical procedures, method of moments, method of maximum likelihood, method of least squares etc. Two most commonly used methods namely the method of moments and the method of maximum likelihood will now be discussed. The graphical method will be presented in conjunction with the topic on probability plotting.

Method of moments. This method simply makes use of the relationships between the parameters and the moments of the distribution. For a distribution with k number of parameters we need to compute the first k moments ; may be about the origin or may be the central moments. Thus, for example, if the assumed distribution is log-normal, its two parameters μ_n and σ_n are related to the mean and variance of the distribution μ_x and σ_x^2 as

$$\mu_n = \frac{1}{2}\ln\left(\frac{\mu_x^4}{\mu_x^2 + \sigma_x^2}\right) \text{ and } \sigma_x^2 = \ln\left(\frac{\sigma_x^2 + \mu_x^2}{\mu_x^2}\right)$$

Taking $\bar{x}$ and s_x to be the estimates of μ_x and σ_x respectively, the estimates of μ_n and σ_n namely $\hat{\mu}_n$ and $\hat{\sigma}_n$ can be obtained as

$$\hat{\mu}_n = \frac{1}{2}\ln\left(\frac{\bar{x}^4}{\bar{x}^2 + s_x^2}\right) \quad \text{...(6.10)}$$

$$\hat{\sigma}_n^2 = \ln\left(\frac{s_x^2 + \bar{x}^2}{\bar{x}^2}\right) \quad \text{...(6.11)}$$

In the case of two parameter gamma distribution, using the relations between its parameters and the moments given by Eq. (4.27) and Eq. (4.28), $\hat{\eta}$ and $\hat{\lambda}$ the estimators of η and λ respectively can be obtained from

$$\hat{\eta} = \left(\frac{\bar{x}}{s_x}\right)^2 \quad \text{...(6.12)}$$

$$\hat{\lambda} = \frac{\bar{x}}{s_x^2} \quad \text{...(6.13)}$$

In the case of Gumbel's distribution, for instance, using Eqs. (4.38) and (4.39), the parameter estimates can be obtained from

$$\hat{\alpha} = 1.28255/s_x \quad \text{...(6.14)}$$

$$\hat{\beta} = \bar{x} - 0.45005\, s_x \quad \text{...(6.15)}$$

Method of maximum likelihood. In this method, the estimators are choosen to be those which will maximise the likelihood function using the observed sample data. Let $f(x, \alpha, \beta, ...)$ be the probability density function of the random variable X, whose parameters $\alpha, \beta, ...$ are to be estimated. Let $(x_1, x_2, ..., x_n)$ be the observed sample of X. Then the product $L = f(x_1, \alpha, \beta, ...). f(x_2, \alpha, \beta, ...). ...... f(x_n, \alpha, \beta, ...)$ is called the likelihood function of a sample of size n. It can also be written as

$$L(x, \alpha, \beta, ...) = \prod_{i=1}^{n} f(x_i, \alpha, \beta, ...) \qquad ...(6.16)$$

Now the parameter estimation procedure is to find the values of $\alpha, \beta, ...$ which will maximise the likelihood function. This can be done by taking the partial derivatives of $L(x, \alpha, \beta, ...)$ with respect to $\alpha, \beta, ...$ and equating the resulting expressions to zero. These equations, as many in number as the parameters, are then solved simultaneously for the unknown parameters. Since many probability distributions involve the exponential function, it is many times easier to maximise the natural logarithm of the likelihood function. Since the logarithmic function is monotonic, the values of $\alpha, \beta, ...$ that maximum the logarithm of the likelihood function will also maximise the likelihood function.

$$\ln [L(x, \alpha, \beta, ...)] = \ln \left[\prod_{i=1}^{n} f(x_i, \alpha, \beta, ...)\right]$$

$$= \sum_{i=1}^{n} \ln [f(x_i, \alpha, \beta, ...)]$$

Now, $\ln [L(x, \alpha, \beta, ...)]$ is maximised when

$$\frac{\partial}{\partial \alpha} \sum_{i=1}^{n} \ln [f(x_i, \alpha, \beta, ...)] = 0$$

$$\frac{\partial}{\partial \beta} \sum_{i=1}^{n} \ln [f(x_i, \alpha, \beta, ...)] = 0$$

$$\vdots \qquad \vdots \qquad \vdots$$

The solution of the above equations would yield the maximum likelihood estimates. The estimates obtained by the maximum likelihood method are the best in a statistical sense. The practical application of this method is often very complicated requiring trial and error solutions.

Example 6.1. *Derive the expressions for the maximum likelihood estimators for the parameters of normal distribution.*

Sol. For a normal distribution with parameters μ and σ

$$f(x) = \frac{1}{\sigma \sqrt{2\pi}} e^{-(x-\mu)^2/2\sigma^2}$$

$$L(x, \mu, \sigma) = f(x_1, \mu, \sigma) \,.\, f(x_2, \mu, \sigma) \;......\; f(x_n, \mu, \sigma)$$

$$= \frac{1}{\sigma^n (2\pi)^{n/2}} e^{-\sum_{i=1}^{n} (x_i - \mu)^2 / 2\sigma^2}$$

$$\ln \; [L(x, \mu, \sigma)] = -n \ln \sigma - \frac{n}{2} \ln (2\pi) - \frac{1}{2\sigma^2} \sum_{i=1}^{n} (x_i - \mu)^2.$$

$$\frac{\partial}{\partial \mu} \ln [L(x, \mu, \sigma)] = -\frac{1}{2\sigma^2} \sum_{i=1}^{n} 2(x_i - \mu)(-1) = 0$$

$$\sum_{i=1}^{n} (x_i - \mu) = 0 \; ; \text{ or } \; \mu = \frac{1}{n} \sum_{i=1}^{n} x_i.$$

$$\frac{\partial}{\partial \sigma} \ln [L(x, \mu, \sigma)] = \frac{-n}{\sigma} + \frac{1}{\sigma^3} \sum_{i=1}^{n} (x_i - \mu)^2 = 0$$

$$\frac{n}{\sigma} = \frac{1}{\sigma^3} \sum_{i=1}^{n} (x_i - \mu)^2 \; ; \text{or } \sigma^2 = \frac{1}{n} \sum_{i=1}^{n} (x_i - \mu)^2.$$

Thus we see that the estimators of method of maximum likelihood for the parameters of the normal distribution are the same as those given by the method of moments. This result is only incidental and not general.

6.3. SELECTION OF DISTRIBUTION

Fitting an appropriate probability distribution involves three steps namely (*i*) the selection of a distribution, (*ii*) the estimation of its parameters and (*iii*) testing its goodness of fit to the observed data. Regarding the selection of proper distribution underlying the process there are no hard and fast rules. One may be tempted to conclude that a proper procedure for selection of a distribution could be to consider a wide variety of distribution functions that are described in Chapters 4 and 5, estimate their parameters using the methods described in Section 6.2, evaluate their adequacy using the testing procedures to be discussed in Sections 6.5 and 6.6 and pass one of them which provides the best fit to the data as the correct choice. However, no such approach is being suggested here.

The best probability distribution to be used to fit the observed data cannot be determined analytically. Whenever possible the selection of the distribution should be based on an understanding of the underlying physical process. Certain limit theorems such as the central limit theorem and the asymptotic theory of extreme values might provide some guidance. For example, one can think of normal distribution for annual rainfall or annual runoff since this variable is the sum of 365 daily values and the justification lies in central limit theorem. Similarly, the extreme value distribution is the most appropriate choice for annual peak flood or annual peak rainfall. The past experience regarding the use of various distributions should be evaluated and used in screening the distributions for possible application in a particular situation. For example, the range of the variable in the distribution function, the general shape of the distribution, and the descriptors like skewness coefficient and kurtosis coefficient many times indicate whether a particular distribution may or may not be applicable in a given situation. If the sample skewness coefficient is very small (that is $| g_1 | \approx 0$) and the sample kurtosis coefficient is around 3, then it will lead the choice to normal distribution. On the other hand, higher skewness coefficient calls for assymetric distributions such as gamma, log-normal etc. In selecting a particular shape of distribution one may be tempted to select a distribution with a large number of parameters. Generally, the more parameters a distribution has, the better it will adapt to a set of data. However, if the sample size is small the reliability in estimating more than 2 or 3 parameters may be quite low. Thus a compromise must be made between flexibility of the distribution and reliability of the parameters. When two or more distributions appear to describe a given set of data equally well, the distribution that has been traditionally used should be selected unless there are compelling reasons for selecting the other distribution. However, if traditionally used distribution is clearly inferior, its use need not be continued.

Now we shall discuss the techniques of assessing the reasonableness of a selected distribution. Such an assessment is very important because unless the selected distribution is satisfactory it cannot be used to make the predictions. There are two different approaches that are considered to make the assessment. They are (*i*) Probability plotting and (*ii*) Statistical tests.

Probability plotting is a subjective method because in this method whether the assumed distribution is fitting well to the observed data or not is based on visual examination rather than any statistical calculation. This method is very simple and, as mentioned earlier, it provides the estimation of parameters in addition to an evaluation of appropriateness of the selected distribution. The method of probability plotting is described in Section 6.4.

The other category of methods involving statistical tests are objective and provide a probabilistic framework to evaluate the adequacy of the distribution. Thus they are useful in supplementing the probability plots, especially when the plots fail to provide a clear cut decision. The two statistical tests which are commonly employed to judge the reasonableness of the distribution are (*i*) the *Chi-square test* and (*ii*) the *Smirnov-Kolmogorov test*. These tests are discussed in Sections 6.5 and 6.6 respectively.

6.4. PROBABILITY PLOTTING

A probability plot is a plot between the magnitude of the random variable and the probability. The probability may be *cumulative probability i.e.*, $P[X \le x]$, the *exceedence probability*, *i.e.* $P[X \ge x]$, or the return period which is the reciprocal of the exceedence probability. If we know the range of the random variable, then the smallest population value is to be assigned a cumulative probability of 0 and the largest population value is to be assigned a cumulative probability of 1. In a probability plot prepared from the sample data the probability of 0 and 1 should be avoided because one can never be sure that a sample contains the smallest and the largest values of the unknown population. Determining the probability to be assigned to a data value is referred to as determining the plotting position.

Let $x_1', x_2', ..., x_n'$ be the original independent observations that are recorded in the same chronological sequence of their occurrence. Let $x_1, x_2, ..., x_n$ be the same observations arranged in the descending order of magnitude such that $x_1 \ge x_2 \ge x_3 \ge ...... \ge x_m \ge ...... \ge x_n$. The general suffix m in the ordered sequence of observations is called the *rank* or the *order* of the observation. This means that the highest value among the observations gets a rank of 1, the next highest gets a rank of 2, and so on while the observation with the least value gets a rank equal to the size of the sample n. When a particular value occurs more than once, all of them should be given the same rank as they are all equal. For example, if a value of 670 occurs twice and is placed in 8th and 9th positions in the descending order then the value 670 is assigned a rank of 9. There are various formulae suggested for obtaining the plotting positions of each of the observations on a probability paper. These formulae give either the cumulative probability or the exceedence probability (or its reciprocal the return period) for any of the ordered observation in terms of its rank m and the sample size n. The return period will be expressed in the same time units as the sampling interval. For example, if we are analysing the annual peak floods then the return period will be in years. If we are analysis the monthly discharges, then the return period will be in months. Ultimately the return period will be expressed only in years after proper conversion. Some of the plotting position formulae which are commonly used in hydrology are given below. In all these formulae $P(x_m)$ indicates the cumulative probability, that is $P[X \le x_m]$.

California formula

$$\text{Exceedence probability} = 1 - P(x_m) = \frac{m}{n} \qquad ...(6.17)$$

$$\text{Return period} \qquad T = \frac{n}{m} \qquad ...(6.18)$$

Hazen's formula

$$1 - P(x_m) = \frac{m - 0.5}{n} = \frac{2m - 1}{2n} \qquad ...(6.19)$$

$$T = \frac{2n}{2m - 1} \qquad ...(6.20)$$

Weibull's formula

$$1 - P(x_m) = \frac{m}{n + 1} \qquad ...(6.21)$$

$$T = \frac{n + 1}{m} \qquad ...(6.22)$$

Consider an observation x_m with rank m. The observed data then reveals that this value x_m has been equalled or exceeded m times out of n observations. Therefore from the frequency interpretation one can say that the probability of x_m being equalled or exceeded is m/n and this is the interpretation of the California formula. But the disadvantage of this formula given by Eq. (6.17) is that it gives an exceedence probability of 1 when $m = n$, which should not appear in a sample of finite size. Hazen's formula tries to avoid this disadvantage by defining the exceedence probability as $(m - 0.5)/n$. But when $m = 1$, the Hazen's formula gives the return period of $2n$ which is double the sample size. The Weibull's plotting position formula given by Eq. (6.21) which will not assign a cumulative probability of either 0 or 1 and which gives a maximum return period of only $(n + 1)$ is therefore most commonly used in hydrology.

There are other plotting position formulae such as Chegodayev's, Blom's, Turkey's and Gringorten's. Most of the plotting position formulae are represented by the following equation.

$$1 - P(x_m) = \frac{m - b}{n + 1 - 2b} \qquad ...(6.23)$$

For example, $b = 0.5$, for Hazen's formula, $b = 0.3$ for Chegodayev's formula, $b = 0$ for Weibull's formula, $b = 0.375$ for

Blom's formula, $b = 1/3$ for Turkeys formula and $b = 0.44$ for Gringorten's formula.

After obtaining the plotting positions for various ordered values of the sample using one of the above formulae, if a plot is made between the value of the random variable x_m as the abscissa and its cumulative probability $P(x_m)$ [or the exceedence probability $1-P(x_m)$ or the return period T] as the ordinate, the resulting graph is called a *distribution graph*, or a *probability plot*, or an *empirical distribution*. Sometimes a plot between x_m and the return period T may be more convenient, especially when we are dealing with annual peak events like floods. The probability plot made on an ordinary graph may be extrapolated or interpolated to obtain the probabilities for any given magnitude of the random variable or *vice-versa* even though we do not know anything about the probability distribution followed by the random variable under consideration. Results obtained from such extrapolation are usually influenced by few end points and also the results obtained may differ considerably from person to person.

Now if the plot is made on a *probability paper* of an assumed distribution instead of on an ordinary graph sheet, and if a straight line appears to be a good fit for the plotted points then the distribution may be taken to be the right choice to model the random variable being analysed. Incidentally, the probability plot also provides a means of estimating the parameters of the distribution.

The probability paper of any distribution is so designed that the cumulative distribution function plots as a straight line on this paper. To achieve this the ordinary scale representing the cumulative probability will have to be converted into a special scale while the abscissa scale representing the random variable remains to be an ordinary scale. The probability papers for important distributions like normal, binomial, and Gumbel's extreme value distribution are commerically available. Even if they are not available, they can be very easily prepared. Consider, for example, the normal distribution. Its cummulative distribution function when plotted on an ordinary graph will be as shown in Fig. 6.1 (*a*). The same can be transformed to plot as a straight line by adjusting the vertical scale in accordance with known coordinates of certain points as given in Table 6.1. Thus the vertical scale will be so marked that all the points such as (– 3.0, 0.0013), (– 2.0, 0.0227), (– 1.00, 0.1587), (0.0, 0.50), (1.0, 0.8413), (2.0, 0.9772), (3.0, 0.9987) will exactly lie on a straight line (of any slope). Such a construction is shown in Fig. 6.1 (*b*). If the straight line and the broken lines used in fixing the vertical

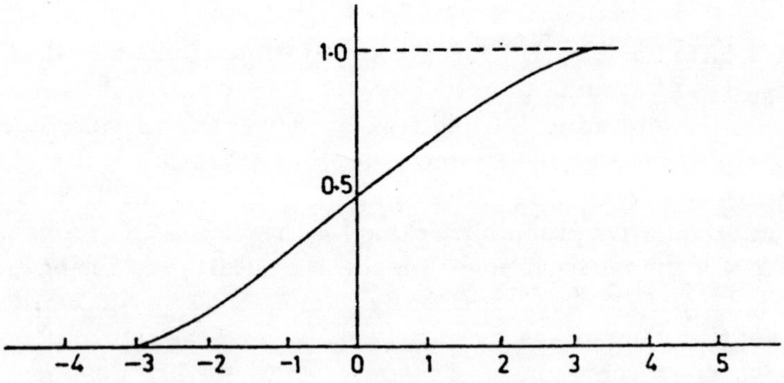

(a) CDF of normal distribution—ordinary plot

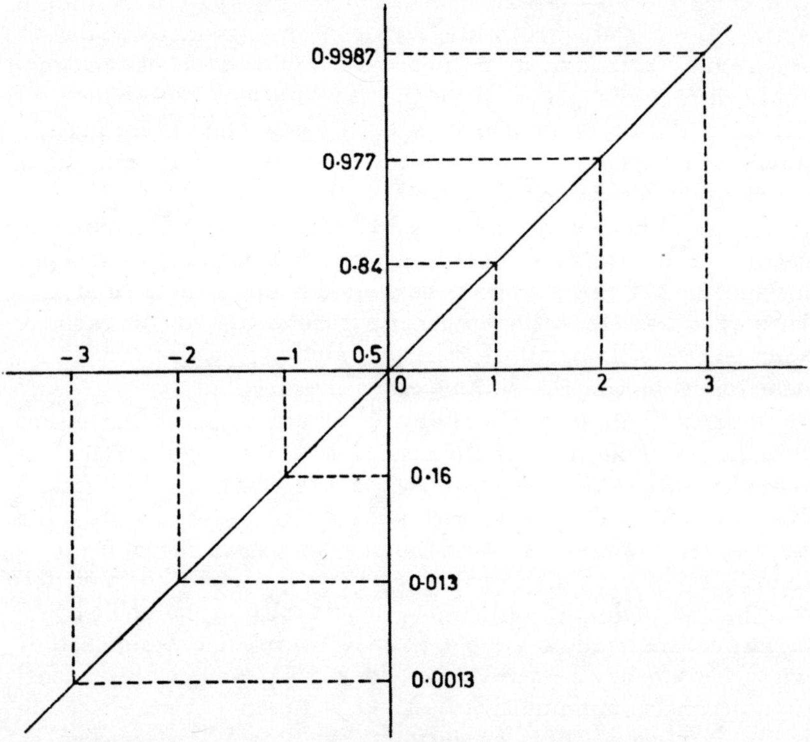

(b) CDF of normal distribution – transformed plot

Fig. 6.1. Normal probability paper.

scale are removed retaining the probability scale on the vertical axis, Fig. 6.1 (*b*) can serve as a normal probability paper. The scale on the abscissa scale can be appropriately fixed to accommodate the range of the observed sample. Probability plot of a near straight line shape on normal probability paper indicates the normal distribution as the right choice.

Table 6.1. Coordinates for cumulative standardised normal distribution

z	$F(z)$	z	$F(z)$
– 3	0.0013	0.5	0.6915
– 2.5	0.0062	1.0	0.8413
– 2.0	0.0027	1.5	0.9332
– 1.5	0.0668	2.0	0.9772
– 1.0	0.1587	2.5	0.9938
– 0.5	0.3085	3.0	0.9987
0	0.5000		

To prepare a probability paper for a log-normal distribution, the vertical scale prepared as described above remains the same and the horizontal scale will be made logarithmic with the required number of cycles. Alternatively, the normal probability paper itself can be used to verify the choice of a log-normal distribution by plotting the logarithm of x_m instead of x_m itself *versus* its cumulative probability. Such a plot appearing as an approximate straight line will lead to the conclusion that a lognormal distribution is a fairly good fit for the random variable being analysed.

Suppose that the observed data plots as an approximate straight line on a normal probability paper confirming that the random variable under consideration can be represented by normal distribution. Then, after fitting a straight line by eye to the plotted points the mean of the distribution can be obtained as the value corresponding to a cumulative probability of 0.5 since we known that the normal distribution is symmetrical about its mean. Again we know from Table 6.1 that in the case of normal distribution the values of the variable corresponding to cumulative probabilities of 0.1587 and 0.8413 are $(\mu - \sigma)$ and $(\mu + \sigma)$ respectively. By reading the values of the random variable against these two cumulative probabilities and dividing the difference of the values by 2, one can get an estimate of the standard deviation. Example 6.2 demonstrates the use of normal probability paper in verifying the choice of normal distribution and also the graphical method of estimating the parameters.

When the adequacy of the Gumbel's distribution is to be assessed the probability plot can be made on *Powell's probability paper*, once the data values are ranked and plotting positions computed. The Powell's probability paper is so designed that the value of the random variable following Gumbel's distribution plotted on this paper against its cumulative probability will be a straight line. The same result can also be achieved by plotting the reduced variate y against the random variable x_m on an ordinary graph sheet. A straight line obtained for this plotting indicates that the extreme value distribution is a correct choice. The slope of the line fitted to the points may be taken to be an estimate of the parameter α, while the intercept on horizontal axis corresponding to $y = 0$ may be taken to be an estimate of β. Example 6.3 illustrates the probability plotting for Gumbel's distribution and the graphical method of estimating its parameters.

Example 6.2. *Annual runoff of a river in units of tmcft (thousand million cubic feet) for a period of 20 years from 1951 to 1970 are given below. Verify the reasonableness of normal and log-normal distributions to fit to this data. Also estimate graphically the parameters of the normal distribution if it were to be an adequate choice.*

Sol. For convenience, the computation of plotting positions is presented in a tabular form in Table 6.2.

Table 6.2. The plotting positions for Example 6.2

Year	*Annual runoff* x^1	*Runoff in descending order* x	*Rank* m	*Exceedence Probability* $1 - P(x_m) = m/(n+1)$	*Cumulative Probability* $P(x_m)$	$\ln x$
(1)	*(2)*	*(3)*	*(4)*	*(5)*	*(6)*	*(7)*
1951	136.1	446.8	1	0.048	0.952	6.102
1952	73.0	382.9	2	0.095	0.905	5.947
1953	169.4	293.0	3	0.143	0.857	5.680
1954	235.4	257.2	4	0.190	0.810	5.550
1955	446.8	244.4	5	0.238	0.762	5.500
1956	382.9	242.8	6	0.286	0.714	5.492
1957	196.8	235.4	7	0.333	0.667	5.461
1958	257.2	221.2	8	0.381	0.619	5.400
1959	221.2	218.3	9	0.429	0.571	5.386
1960	82.8	200.1	10	0.476	0.524	5.300
1961	196.3	196.8	11	0.524	0.476	5.282
1962	293.0	196.3	12	0.571	0.429	5.280

1963	242.8	169.4	13	0.619	0.381	5.132
1964	218.3	136.1	14	0.667	0.333	4.913
1965	98.5	100.1	15	0.714	0.286	4.606
1966	100.1	98.5	16	0.762	0.238	4.590
1967	200.1	86.2	17	0.809	0.191	4.457
1968	68.5	82.8	18	0.857	0.143	4.416
1969	86,2	73.0	19	0.905	0.095	4.290
1970	244.4	68.5	20	0.952	0.048	4.227

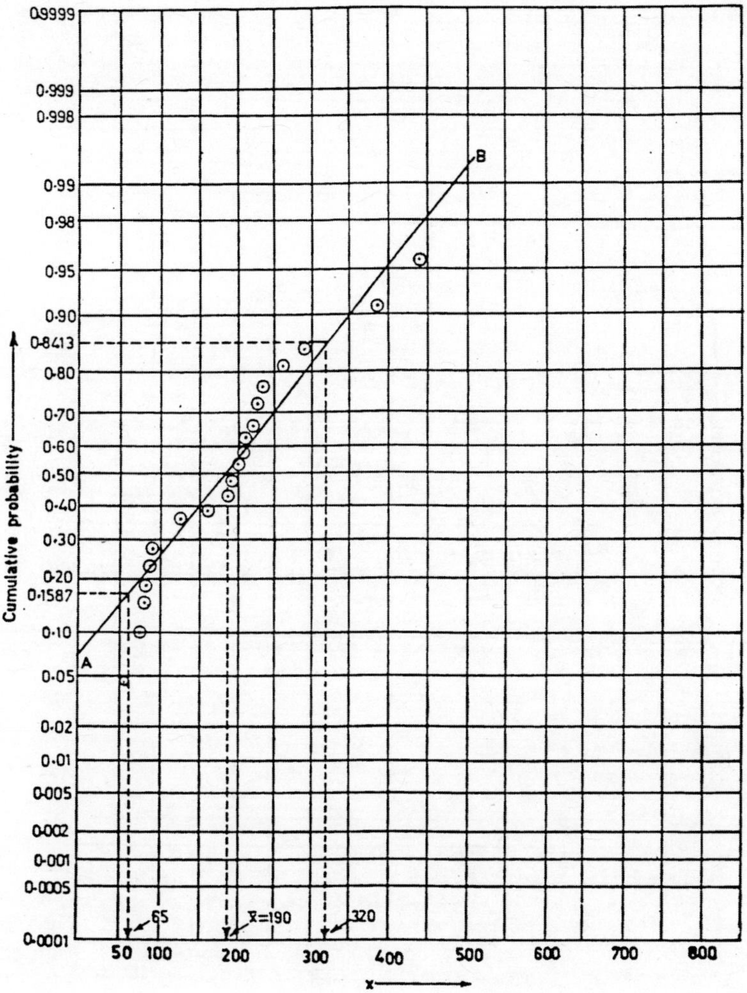

Fig. 6.2. Probability plot for normal distribution.

Columns (1) and (2) show the year and the annual runoff in that year denoted by x^1. In column (3) the runoff values, x, arranged in descending order are shown and the ranks assigned to these values, m, are entered in column (4). column (5) shows the exceedence probability $[1 - P(x_m)]$ computed using Eq. (6.21). The cummulating probabilities obtained as (1 – exceedence probability) are given in column (6) while column (7) gives the logarithm of x_m taken to the natural base.

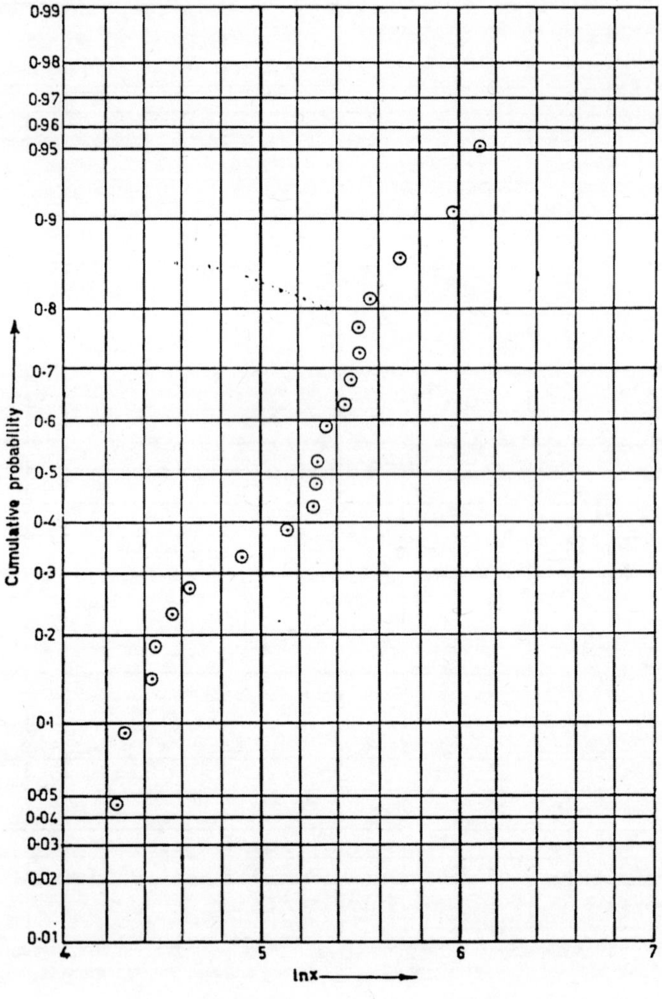

Fig. 6.3. Probability plot for log-normal distribution.

The ordered annual runoff of column (3) *versus* its cumulative probability of column (6) are plotted on a normal probability paper as shown in Fig. 6.2. The points do not reveal a clearcut straight line. Nevertheless, since we are dealing with annual runoff, central limit theorem is in favour of normal distribution. Therefore a straight line AB is eye fitted to the plotted points for the purpose of estimating the parameters. As pointed out earlier, $\bar{x}$ is the value corresponding to a cumulative probability of 0.5 which in this case is 190.0. The values corresponding to cumulative probabilities of 0.8413 and 0.1587 are 320 and 65, whose difference is 255. Hence s_x = 255/2 = 127.5. Thus the graphically estimated mean and standard deviation are 190 and 127.5 tmcft respectively.

If the parameters of this data are estimated using the method of moments with Eqs. (6.2) and (6.6) one gets $\bar{x}$ = 197.5 tmcft and s_x = 101.94 tmcft.

The logarithm of annual runoff given in column (7) is next plotted against the cumulative probability of column (6) on a normal probability paper in Fig. 6.3. This plot also does not point to a clearcut straight line. However, going by central limit theorem, the normal distribution may be favoured.

Example 6.3. *The annual peak discharges in m^3/s of a stream for a period of 20 years are given below. Verify whether the Gumbel's distribution is a correct choice to describe the annual peak discharge.*

1340, 1380, 1450, 618, 523, 508, 1220, 1780, 1060, 412, 184, 1480, 876, 113, 516, 1090, 944, 397, 282, 353.

Sol. We have that exceedence probability = $1 - P(x_m) = 1/T$

Therefore the cumulative probability and the return period are related by

$$P(x_m) = (1 - 1/T).$$ Also for Gumbel's distribution

$$P(x_m) = e^{-e^{-y}} \quad \text{or} \quad y = -\ln [\{-\ln \{P(x_m)\}].$$

In Table 6.3 column (1) shows the observed annual peak discharges x' in the chronological order of their occurrence. The same discharges are arranged in descending order of magnitude (denoted by x) and are entered in column (2), while the ranks assigned to them are shown in column (3). The return period T computed from Eq. (6.22) are entered in column (4) and the cumulative probability computed as $P(x_m) = (1 - 1/T)$ is given in column (5) and the corresponding reduced variate y obtained from $y = -\ln [-\ln \{P(x_m)\}]$ is shown in column (6).

Table 6.3. The plotting positions for Example 6.3

x' (1)	x (2)	m (3)	$T = \frac{n+1}{m}$ (4)	$P(x_m) = 1 - 1/T$ (5)	$y = -\ln[-\ln\{P(x_m)\}]$ (6)
1340	1780	1	21.00	0.9524	3.0206
1380	1480	2	10.50	0.9048	2.3022
1450	1450	3	7.00	0.8571	1.8695
618	1380	4	5.25	0.8095	1.5543
523	1340	5	4.20	0.7619	1.3022
508	1220	6	3.50	0.7143	1.0893
1220	1090	7	3.00	0.6667	0.9027
1780	1060	8	2.63	0.6190	0.7347
1060	944	9	2.33	0.5714	0.5804
412	876	10	2.10	0.5238	0.4359
184	618	11	1.90	0.4762	0.2895
1480	523	12	1.75	0.4286	0.1658
876	516	13	1.62	0.3809	0.0354
113	508	14	1.50	0.3333	– 0.0940
516	412	15	1.40	0.2857	– 0.2254
1090	397	16	1.31	0.2380	– 0.3615
944	353	17	1.24	0.1905	– 0.5057
397	282	18	1.17	0.1429	– 0.6656
282	184	19	1.11	0.0952	– 0.8552
353	113	20	1.05	0.0476	– 1.1135

Now a graph is plotted between the reduced variate y of column (6) on the vertical axis and the value of annual peak x of column (2) on horizontal axis as shown in Fig. 6.4. This plot indicates that the points are more or less clustered around a straight line and therefore the Gumbel's distribution may be taken to be adequate to represent the annual peak flows. A straight line such as AB fitted by eye can be extrapolated or interpolated to obtain the required answers regarding the probabilities of peak discharges.

Now $\hat{\alpha}$, an estimate of the parameter α may be obtained as the slope of the line AB, since AB represents the equation $y = \alpha(x - \beta)$. The value of $\hat{\alpha}$ in the present case is (1/415). Similarly $\hat{\beta}$ the estimate of β is equal to the intercept on the horizontal axis where $y = 0$, which is 590 m^3/s.

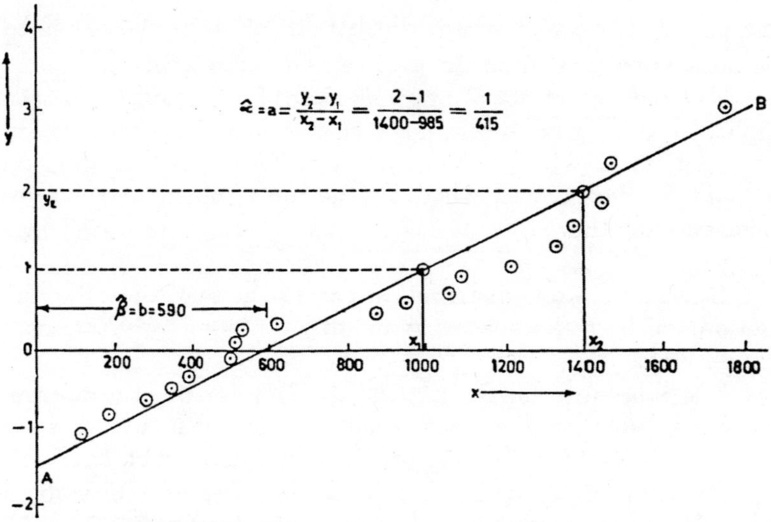

Fig. 6.4. Probability plot for Gumbel's distribution.

6.5. CHI-SQUARED TEST

Let the range of a continuous random variable be divided into k class intervals which are mutually exclusive. Let n_i be the number of observations falling in the ith class interval where $i = 1, 2, ..., k$. Then the total number of observations $n = n_1 + n_2 + ... + n_k$. Select a probability distribution whose adequacy is to be tested. Estimate the parameters of the distribution either by the method of moments or by the method of maximum likelihood. Using the probability distribution determine the probability p_i with which the random variable lies in the each class interval i. Then the expected number of observations in any class interval, e_i, may be obtained as $e_i = np_i$, now the Chi-square parameter is computed as

$$\chi^2 = \sum_{i=1}^{k} \frac{(n_i - e_i)^2}{e_i} \qquad ...(6.24)$$

This parameter will follow a Chi-square distribution with the number of degrees of freedom equal to $(k - h - 1)$, where h is the number of parameter estimated. Fix 'α' the significance level of the test. Usually this level would be either 10% or 5%, that is α is equal to either 0.1 or 0.05. Obtain the critical value χ_0^2 for the χ^2 distribution with $(k - h - 1)$ degrees of freedom such that $P[\chi^2 \geq \chi_0^2] = \alpha$. If the Chi-square value computed from Eq. (6.24) is less than the critical value χ_0^2, then we will accept the hypothesis that the assumed distribution is a good fit at the given significance level.

The number of class intervals should be at least 5. If the sample size n is large, the number of class intervals may be approximately fixed at [10 + 1.33 ln (n)]. The size of the class interval need not be uniform, but it is better if the non-uniform intervals are so chosen that there are at least 5 observations in each class interval.

If more than one distribution passes the test, then the distribution which gives the least value for χ^2 is taken to be the most appriate choice.

The procedure for testing the goodness of fit of a discrete distribution is exactly the same except that the class intervals are replaced by the k discrete values that the random variable can take.

Example 6.4. *The annual runoff in tmcft of a river for a period of 78 years are given below. Test the goodness of fit of normal distribution at 10% significance level using Chi-square test.*

Years	*1 to 10*	*11 to 20*	*21 to 30*	*31 to 40*	*41 to 50*	*51 to 60*	*61 to 70*	*71 to 78*
	2084	1770	3049	2324	2085	2129	2439	3389
	2182	1272	2374	2281	1927	1960	2969	2063
	2611	1891	3721	2177	1990	2840	4165	1939
	2695	2276	3029	2305	2045	2525	2730	2519
	2587	2472	1007	2212	2613	2311	3113	2116
	1124	2144	2270	1918	2194	2544	3477	2665
	2795	2422	1690	2196	2287	2628	3060	2725
	2057	1451	2164	2903	1715	1970	3755	2137
	1869	1965	2063	2703	2169	1749	3075	
	3160	1809	2471	2936	2333	2919	2751	

Sol. The mean, standard deviation, skewness coefficient and the kurtosis coefficient of the given sample are computed to be 2390.37, 571.31, 0.45 and 3.80 using Eqs. (6.2). (6.4), (6.6), (6.8) and (6.9) respectively. Invoking the central limit theorem, we expect the normal distribution to be a good fit for the annual flows. This is also suggested by a small value of sample skewness coefficient. Though the sample kurtosis coefficient is greater than the kurtosis coefficient for normal distribution (which is 3), this cannot be given much importance because the kurtosis coefficient is based on 4th moment which cannot be estimated reliably unless the sample size is really large.

Table 6.4. Computations of Example 6.4 for Chi-square test on Normal Distribution

i	Class interval	No. of observations n_i	Standardised normal variate z_i of upper limit of class interval	$F(z_i)$	Probabilty of class-interval $p_i = F(z_i) - F(z_{i-1})$	Expected number of observations $e_i = np_i$	$\frac{(n_i - e_i)^2}{e_i}$
(1)	(2)	(3)	(4)	(5)	(6)	(7)	(8)
1	< 1400	3	– 1.73	0.04	0.04	3.12	0.0046
2	1401—1800	5	– 1.03	0.15	0.11	8.58	1.4937
3	1801—2200	26	– 0.33	0.37	0.22	17.16	4.5539
4	2201—2600	18	0.37	0.64	0.27	21.06	0.4446
5	2601—3000	16	1.07	0.86	0.22	17.16	0.0784
6	3001—3400	6	1.77	0.96	0.10	7.8	0.4153
7	3401—3800	3	2.47	0.99	0.03	2.34	0.1861
8	> 3800	1		1.00	0.01	0.78	0.0620
	Σ	78			1.0000	78.00	7.2386

The necessary computations to perform the Chi-square test are given in Table 6.4. The range of observations is divided into 8 class intervals. The lower and upper bound of these class intervals are shown in col. (2). The number of observations falling in each class interval are found out by counting and are entered in col. (3). The standardised normal variate z_i corresponding to the upper limit of each class interval are given in col. (4). For example, the upper limit of fourth class interval is 2600 and the value of corresponding z_i is obtained as $z_i = (2600 - 2390.37)/571.31 = 0.37$. The values of $F(z_i)$ for each z_i are read out from the Table A-1. The values appearing in col. (6), col. (7) and col. (8) are self-explanatory.

The totals of col. (3), col. (6) and col. (7) provide a check on the calculations. The total of col. (8) is the value of the Chi-square parameter which in the present case is 7.2386. The number of class intervals $k = 8$, the number of parameters estimated $h = 2$ (that is $\bar{x}$ and s_x) and therefore the number of degrees of freedom for the Chi-square parameter is $(k - h - 1) = (8 - 2 - 1) = 5$. The critical value χ_0^2 for $\alpha = 0.10$ and 5 degrees of freedom read out from Table A-2 is 9.24.

Since the computed value of χ^2, which is 7.2386, is less than χ_0^2, we accept the hypothesis that the normal distribution is a good fit for the given data.

Example 6.5. *The number of rainy days in the first week of July at a particular raingauge station for a period of 50 years is recorded as given below :*

No. of rainy days :	*0*	*1*	*2*	*3*	*4*	*5*	*6*	*7*
No. of years :	*2*	*9*	*15*	*13*	*7*	*3*	*1*	*0*

Assuming that the Binomial distribution can be used to model this event, estimate the parameter p. Test the goodness of fit of the binomial distribution at 5% significance level by Chi-square test.

Sol. The total number of days on which observations were made is equal to 50 year × 7 days in each year = 350 days. The number of days on which the rainfall is observed at the station = (2 × 0) + (9 × 1) + (15 × 2) + (13 × 3) + (7 × 4) + (3 × 5) + (1 × 6) + (0 × 7) = 0 + 9 + 30 + 39 + 28 + 15 + 6 + 0 = 127. Therefore the probability for a rainy day = 127/350 = 0.363. That is, $p = 0.363$ and $q = (1 - p) = 0.637$. The other computations required to carry out the Chi-square test are given in Table 6.5.

Table 6.5. Computations for Chi-square test on Binomial distribution

No. of rainy days x_i	*Observed no. of years* n_i	$P[X = x_i] = p_i = \binom{7}{x} P^x \cdot q^{7-x}$	*Expected No. of years* $e_i = 50p_i$	$\frac{(n_i - e_i)^2}{e_i}$
0	2	0.0425	2.13	0.0079
1	9	0.1697	8.49	0.0306
2	15	0.2902	14.51	0.0165
3	13	0.2756	13.78	0.0442
4	7	0.1571	7.85	0.0920
5	3	0.0537	2.68	0.0382
6	1	0.9102	0.51	0.4708
7	0	0.0008	0.04	0.0400
Σ	50	0.9998	49.99	0.7402

No. of parameters estimated $= h = 1$

No. of class intervals $= k = 8$

Degrees of freedom $= 8 - 1 - 1 = 6$

for 6 degrees of freedom from Table A.2. at $\alpha = 0.05$, $\chi_0^2 = 13.6$

Since $\chi^2 = 0.7402 < 13.6$

we accept the hypothesis that the Binomial distribution is a good fit.

6.6. SMIRNOV-KOLMOGOROV TEST

This is an alternative to the Chi-square test. This test can be conducted as follows. Let $x_1, x_2, \ldots\ldots, x_m, \ldots\ldots, x_n$ be the ordered values of the random variable in a sample of size n arranged in descending order of magnitude. Compute the cumulative probability $P(x_m)$ for each of the observations x_m using the Weibull's formula of Eq. (6.21). Assume a distribution whose goodness of fit is to be tested. Estimate its parameters. Then obtain the theoretical cumulative probability $F(x_m)$ for each ordered observation x_m using the assumed distribution. The absolute difference of $P(x_m)$ and $F(x_m)$, that is $| P(x_m) - F(x_m) |$ is computed for each x_m. The Smirnov-Kolmogorov statistic Δ is the largest value of these differences. That is

$$\Delta = \text{maximum of } | P(x_m) - F(x_m) | \qquad \ldots(6.25)$$

Fix the level of significance for the test such as $\alpha = 0.1$ or 0.05. Obtain the critical value of Smirnov-Kolmogorov statistic Δ_0 from Table A-3 in the Appendix which is given as a function of α and the sample size n. If $\Delta < \Delta_0$, accept the hypothesis that the distribution is a good fit, otherwise reject the hypothesis.

The Smirnov-Kolmogorov test has an advantage over the Chi-square test in that it does not lump the data and compare only discrete categories, but it rather compares all the data in an unaltered form. Also it is easier to compute the value of Δ than χ^2. It is more convenient to adopt, especially when the sample size is small.

If more than one distribution pass the test, then the distribution which gives the least value for Δ is taken to be the most appropriate choice.

Example 6.6. *Test the goodness of fit of normal distribution for the data of Example 6.2. by Smirnov-Kolmogorov test at 10% significance level.*

Sol. Table 6.6 shows the computations required to carry out the test. The ordered values of x and the cumulative probability $P(x_m)$ are directly taken from Table 6.2. The value of z for each value of x is calculated as $z = (x - \bar{x})/s_x$, that is $z = (x - 197.49)/101.94$. Then $F(x)$ is nothing but $F(z)$ itself. The values of $F(z)$ are read out from Table A-1. The other computations appearing in Table 6.6 are self-explanatory. Therefore the value of Δ for the given data is 0.12. The critical value Δ_0 obtained from Table A-3 for $n = 20$ and $\alpha = 0.10$ is 0.26.

Since $\Delta < \Delta_0$, we accept the hypothesis that the normal distribution is a good fit.

Table 6.6. Computations of Example 6.6 for Smirnov-Kolmogorov test on normal distribution

x	$P(x)$	Standardised normal variate $z = \frac{x - \bar{x}}{s_x}$	$F(x) = F(z)$	$\lvert F(x) - P(x) \rvert$	Δ
446.8	0.95	2.44	0.99	0.04	
382.9	0.90	1.82	0.97	0.07	
293.0	0.86	0.94	0.83	0.03	
257.2	0.81	0.59	0.72	0.09	
244.4	0.76	0.46	0.68	0.08	
242.8	0.71	0.44	0.67	0.04	
235.4	0.67	0.37	0.64	0.03	
221.2	0.62	0.23	0.59	0.03	
218.3	0.57	0.20	0.58	0.01	
200.1	0.52	0.03	0.51	0.01	
196.8	0.48	0.00	0.50	0.02	
196.3	0.43	– 0.01	0.50	0.07	
169.4	0.38	– 0.28	0.39	0.01	
136.1	0.33	– 0.60	0.27	0.06	
100.1	0.29	– 0.96	0.17	**0.12**	0.12
98.5	0.24	– 0.97	0.17	0.07	
86.2	0.19	– 1.09	0.14	0.05	
82.8	0.14	– 1.13	0.13	0.01	
73.0	0.10	– 1.22	0.11	0.01	
68.5	0.05	– 1.27	0.10	0.05	

The reader may verify that the lognormal distribution also passes the Smirnov-Kolmogorov Test at 10% significance level. But in that case the value of the statistic Δ will be 0.16 which is greater than the value in the present case.

6.7. FREQUENCY FACTORS

The discussion presented so far could be called the *frequency analysis*, since frequency analysis is merely a procedure for estimating the frequency of occurrence or probability of occurrence of past and/or future events. Thus probability plotting may be used with or without any assumed distribution for determining frequencies of events. If no distributional assumptions are made, the investigator merely plots the observed data on any kind of paper (not necessarily probability paper) and uses his best judgement to determine the magnitude of the event, past or future, for various probabilities (or return periods). If a distribution is assumed, analytical methods may be used to obtain the magnitude of events for given probabilities. Before using the analytical method, it is suggested that the reasonableness of the assumed distribution be assessed either by probability plotting or by statistical tests. The analytical method uses the concept of *frequency factors*. Ven Te Chow has proposed that an event x_T, with a return period T can be expressed as

$$x_T = \bar{x} + K_T\, s_x \qquad ...(6.26)$$

where K_T is the frequency factor. For a given distribution a relationship between the frequency factor and the corresponding return period can be determined. This relationship may be expressed as a mathematical equation or in the form of a Table. For a two parameter distributions K_T depends only on the return period. In skewed distributions it varies with coefficient of skewness and it can be affected greatly by the record length n.

The method begins with the calculation of the statistical parameters of the proposed distribution by the method of moments from the given data. For a given return period, the frequency factor can be determined from the $K_T - T$ relationship applicable for the proposed distribution and the magnitude of x_T is then computed using Eq. (6.26).

The theoretical $K_T - T$ relationship for some of the probability distributions used in hydrologic frequency analysis are now described.

Normal distribution. From Eq. (6.26) frequency factor can be expressed as

$$K_T = \frac{x_T - \bar{x}}{s_x} \qquad ...(6.27)$$

This is same as the standardised normal variable z. That is $K_T = z$. Therefore for a given return period T, the value of K_T may be obtained as the value of z such that $F(z) = (1 - 1/T)$ from Table A-1 or alternatively from Eq. (4.10).

For log-normal distribution, the same procedure applies except that it is applied to the logarithms of the variable and its mean and standard deviation. That is

$$y_T = \bar{y} + K_T\, s_y \qquad ....(6.28)$$

$$x_T = \exp\,(y_T) \qquad ...(6.29)$$

where $\bar{y}$ and s_y are the mean and standard deviation of the logarithms taken to the natural base.

Gumbel's distribution. For the Gumbel's distribution the frequency factor is given by

$$K_T = -\left\{0.45 + 0.78 \ln\left[\ln\left(\frac{T}{T-1}\right)\right]\right\} \qquad ...(6.30)$$

The values of K_T computed from Eq. (6.30) correspond to an infinite sample size.

When a finite sample size n is used in the analysis, K_T can be obtained from

$$K_T = -\left\{\bar{y}_n + \ln\left[\ln\left(\frac{T}{T-1}\right)\right]\right\} \Big/ \sigma_n \qquad ...(6.31)$$

where $\bar{y}_n$ and σ_n are called the mean and standard deviation of the *reduced extremes* which depend on the sample size n as given in Table A-4.

Log-Pearson type III distribution. For this distribution to be applied, the logarithmic transformation is applied to the observed x series to yield the y series. That is, $y_i = \log(x_i)$. Usually logarithms are taken to the base 10. Then mean $\bar{y}$, standard deviation s_y and the skewness coefficient g_1 are calculated for the logarithms, that is y series. The frequency factor depends on the return period T and the skewness coefficient g_1 and it may be computed from the following equation.

$$K_T = z + (z^2 - 1)\,k + \frac{1}{3}(z^3 - 6z)\,k^2 - (z^2 - 1)\,k^3 + zk^4 + \frac{k^5}{3} \qquad ...(6.32)$$

where $k = g_1/6$ and z is the standardised normal variable. When the skewness coefficient g_1 is zero Eq. (6.32) gives $K_T = z$. This means log-Pearson type III distribution is equivalent to log-normal distribution when the logarithms of the variable have zero skewness.

For a given return period T, and the computed skewness coefficient g_1, the frequency factor K_T is obtained from Eq. (6.32). When Eq. (6.32) is used, the value of z is obtained from Table A-1 or Eq. (4.10). Now y_T is computed from Eq. (6.28) and finally x_T is obtained as

$$x_T = 10^{y_T} \qquad ...(6.33)$$

Example 6.7. *The annual maximum one hour rainfalls in mm observed over a period of 35 years are given below. Compute the mean and standard deviation of this sample.*

25.0 33.5 18.3 29.5 21.0 24.0 37.6 27.0 38.6 29.0 40.6
33.5 34.5 34.5 31.0 44.7 24.9 16.8 48.8 47.8 40.6 31.5 36.0
56.4 32.5 26.4 32.5 17.3 35.6 29.0 46.7 33.5 33.0 32.0 30.5

Also find the mean and standard deviation of the logarithms of the sample.

Sol. $\bar{x} = \dfrac{(25.0 + 33.5 + \ldots\ldots + 32.0 + 30.5)}{35}$

$= 32.97$ mm

$s_x^2 = [(25.0 - 32.97)^2 + (33.5 - 32.97)^2 + \ldots\ldots + (30.5 - 32.97)^2]/(35 - 1)$

$= 80.80$

$\therefore \quad s_x = 8.99$ mm

Taking the logarithms to the natural base, the data gets transformed into

3.22	3.51	2.91	3.38	3.04	3.18	3.63	3.30	3.65	3.37
3.70	3.51	3.54	3.54	3.43	3.80	3.21	2.82	3.89	3.87
3.70	3.45	3.58	4.03	3.48	3.27	3.48	2.85	3.57	3.37
3.84	3.51	3.50	3.47	3.42					

This gives $\bar{y} = 3.4577$ and $s_y = 0.2825$.

Example 6.8. *For the data of Example 6.7, find the one hour annual maximum rainfall with a 50 year return period assuming (a) normal distribution (b) lognormal distribution and (c) Gumbel's distribution.*

Sol. (*a*) For a return period of 50 years, the cumulative probability is $(1 - 1/50) = 0.98$. From Table A-2, when $F(z) = 0.98$ $z = 2.054$.

$$x_T = \bar{x} + K_T s_x$$

$$x_{50} = 32.97 + 2.054 \times 8.99 = 51.44 \text{ mm.}$$

(*b*)

$$y_T = \bar{y} + K_T s_y$$

$$y_{50} = 3.4577 + 2.054 \times 0.2825 = 4.038$$

$$x_{50} = \exp(y_{50}) = e^{4.038} = 56.71 \text{ mm}$$

(*c*) From Table A-4, for $n = 35$, we get $\bar{y}_n = 0.54034$ and $\sigma_n = 1.12847$

$$K_T = -\left\{\bar{y}_n + \ln\left[\ln\left(\frac{T}{T-1}\right)\right]\right\} \Big/ \sigma_n$$

$$K_{50} = -\left\{0.54034 + \ln\left[\ln\left(\frac{50}{49}\right)\right]\right\} \Big/ 1.12847$$

$$= -\{0.54034 - 3.90194\}/1.12847$$

$$= 2.9789$$

$$x_{50} = 32.97 + 2.9789 \times 8.99 = 59.75 \text{ mm.}$$

It may be noted here that the largest value in the sample is 56.4 mm. This corresponds to a return period of 36 years, when return period is computed from Weibull's formula. Therefore, we expect the 50 year rainfall to be greater than 56.4 mm. Thus the answer obtained by using Gumbel's distribution is more acceptable.

Example 6.9. *Annual peak floods of a stream measured in* m^3/s *for a period of 40 years are analysed and the mean, standard deviation and skewness coefficient of logarithms taken to base 10 of these floods are found to be 2.83, 0.138 and + 0.60 respectively. Estimate the 50 year flood using log-Pearson type III distribution.*

Sol. From Table A-1 for $T = 50$ years, $z = 2.054$.

$k = g_1/6 = 0.6/6 = 0.10$; From Eq. (6.32), with $z = 2.054$ and $k = 0.1$, we get $K_{50} = 2.361$.

$$y_T = \bar{y} + K_T\, s_y$$

$$y_{50} = 2.83 + 2.361 \times 0.138 = 3.1558$$

$$x_{50} = 10^{3.1558} = 1431.6 \text{ m}^3/\text{s}.$$

EXERCISES

6.1. Find the maximum likelihood estimator for the parameter λ of the exponential distribution, whose probability density function is given by $f(x) = \lambda e^{-\lambda x}$, for $x \geq 0$.

6.2. Derive the expressions for the maximum likelihood estimators for the parameters α and β of the Gumbel's distribution.

6.3. Show that the maximum likelihood estimators for the parameters μ and σ of normal distribution are the same as those given by the method of moments.

6.4. The annual rainfall over a basin for a period of 21 years has a mean of 2396 mm and a standard deviation of 868 mm and has the following frequency distribution.

Class interval	< 1500	1500 to 2000	2001 to 2200	2201 to 2500	2501 to 3000	3001 to 3500	> 3500
No. of years of occurrence	4	2	5	3	1	3	3

Test the goodness of fit of normal distribution to this data at 5% significance level.

6.5. Given below are the annual peak discharges of a river basin for a period of 19 years. Construct the empirical distribution curve using Weibull's plotting position formula and estimate the flood with a return period of 50 years. Also find the median of the floods and compare it with the mean value of the floods.

2860	3090	4050	1220	1630	1840	450	1610	650	1630
430	360	510	1000	350	630	1450	2670	5140	

6.6. Obtain the expression for the estimators of the parameters of log-normal distribution by the method of maximum likelihood.

6.7. Explain how you would test the goodness of fit of a distribution to the observed data by Chi-square test.

6.8. From any publication of hydrologic data select a river with monthly flows at least for a period of 39 years. From this data take the November discharges and sort them in descending order of magnitude. Compute the plotting positions using Weibulli's formula. Plot the empirical distribution on normal probability paper and also on log-normal probability paper. Fit straightlines by eye and estimate the parameters graphically in each case.

6.9. The mean and standard deviation of annual peak floods of a river estimated from 49 year record are 9800 m^3/s and 2950 m^3/s respectively. Assuming Gumbel's distribution, determine the 100 year and 200 year floods for this river. Use frequency factors ($\bar{y}_n = 0.5481, \sigma_n = 1.1590$).

6.10. Explain how do you test the goodness of fit of a distribution to the observed data by Smirnov-Kolmogorov test.

6.11. For the data of Exercise 6.5, assess whether the Gumbel's distribution is a good fit by Smirnov-Kolmogorov test at 10% significance level.

6.12. The following statistics are obtained from a sample of 150 of a random variable X.

$\bar{x} = 278$; $s_x = 25$; skewness coefficient $g_1 = -0.023$ and kurtosis coefficient $g_2 = 3.084$.

What distribution, in your opinion, could be a correct choice to describe the random variable. Give reasons.

6.13. The mean and standard deviation of a sample of size 20 are computed as 25 and 10 respectively. Later it is found that two more observations with values 35 and 15 have not been included in the calculations. What are the new values of mean and standard deviation if these observations are also included in the calculations ?

6.14. Find the annual maximum one hour rainfall of 50 years return period assuming that log-Pearson type III distribution is a reasonable fit to the data of Example 6.7.

6.15. Plot the data of Example 6.7 on normal probability paper, log-normal probability paper, and Powell's probability paper and comment on the results.

6.16. Test the goodness of fit of normal distribution and log-normal distribution to the data of Exercise 6.5 at 10% significance level by

Smirnov-Kolmogorov test. How do you compare these results with the one obtained in Exercise 6.11.

6.17. If the discharge in a river in excess of certain magnitude lasts continuously for a period of 48 hours, then it is treated as a flood. In a 100 year period, the following number of floods were recorded at a specified location. Assuming that this data can be described by a Binomial distribution estimate the parameter p. Test the goodness of fit by Chi-square test at 10% significance level.

No. of floods in a year	0	1	2	3	4	5	6
No. of occurrences	52	28	12	12	2	1	0

6.18. Analysis of logarithms (taken to the base 10) of annual flood peaks measured in m^3/s of a river yielded the following statistics. $\bar{y} = 3.5$; $s_y = 0.95$ and $g_1 = -0.1$. Determine the 50 year flood using the log-Pearson type III distribution.

6.19. Annual flood peak, expressed in m^3/s, at a particular site on a river follow a zero-skew log-Pearson type III distribution. If the mean and standard deviation of logarithms (to the base 10) of the annual floods are 2.95 and 1.0 respectively, determine the magnitude of the 50 years flood.

6.20. The annual flood peak at a particular site on a stream is modelled by the Gumbel's distribution. If the estimate of 50 year and 100 year floods are 2400 m^3/s and 2730 m^3/s respectively, what is the estimate of 200 year flood ?

7

Correlation and Regression

7.1. CORRELATION

In Chapter 3, we defined the population correlation coefficient between two random variables X and Y in terms of the covariance of X and Y and the variances of X and Y

$$\rho = \frac{\text{cov}(X, Y)}{\sigma_x \cdot \sigma_y} \quad ...(7.1)$$

The sample correlation coefficient r is similarly defined by

$$r = \frac{s_{x,y}}{s_x \cdot s_y} \quad ...(7.2)$$

where $s_{x,y}$ is the sample covariance between X and Y and s_x and s_y are the sample standard deviations of X and Y respectively. If (x_i, y_i) is the ith observation of X and Y and $\bar{x}$ and $\bar{y}$ are the sample means of X and Y, then r can be computed from

$$r = \frac{\frac{1}{n}\sum_{i=1}^{n}(x_i - \bar{x})\ (y_i - \bar{y})}{s_x \cdot s_y} \quad ...(7.3)$$

A careful examination of Eq. (7.3) reveals that the correlation coefficient becomes positive when the numerator is positive and becomes negative when the numerator is negative. Again the numerator becomes positive when the products in the summation are positive most of the times and becomes negative when the products in the summation are negative most of the times. Any product in the summation is positive when x_i which is more than $\bar{x}$ is associated with y_i which is also more than $\bar{y}$ or when x_i which is less than $\bar{x}$ is associated with y_i which is also less than $\bar{y}$. In other words two variables are likely to be positively correlated when higher values of one variable are associated with correspondingly higher values of the other variable and lower values of one variable are associated with correspondingly lower values of the other variable. Similarly, negative correlation between two random variables can be expected when higher values of one variable are associated with the lower values of the other variable and *vice versa*.

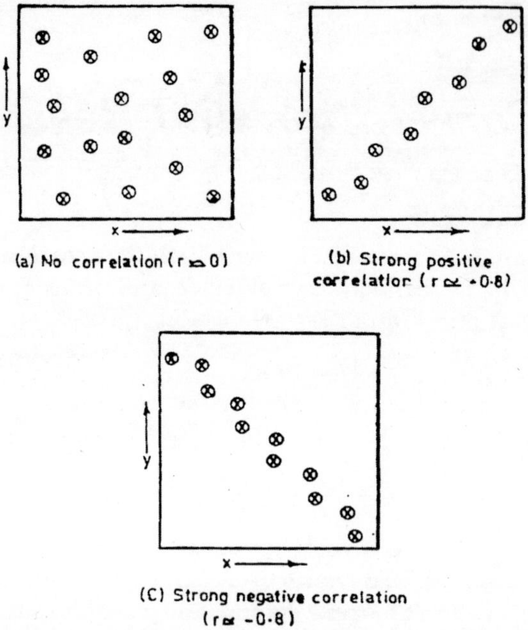

Fig. 7.1. Scatter diagram.

The extent of correlation between the variables can be qualitatively ascertained by preparing a graphical plot between the observed values of the variables. Such a plot is known as a *scatter diagram*. The scatter diagram for the cases of no correlation, strong positive correlation and strong negative correlation are shown in Fig. 7.1. In Fig. 7.1 (*a*), the plotted points are really scattered indicating no particular trend between the variables. In Fig. 7.1 (*b*) though the points are reflecting some scatter, the points are clustered around a seemingly straight line with positive slope. Similarly the points of scatter in Fig. 7.1 (*c*) are clustered around a seemingly straight line with a negative slope. It means then that when two variables are positively correlated, the higher values of one variable Y are associated by and large with the higher values of the other variable X. Likewise when the variables are negatively correlated the higher values of Y are associated with lower values of X and lower values of Y are associated with higher values of X. When they are not correlated we do not notice such trends in the scatter diagram.

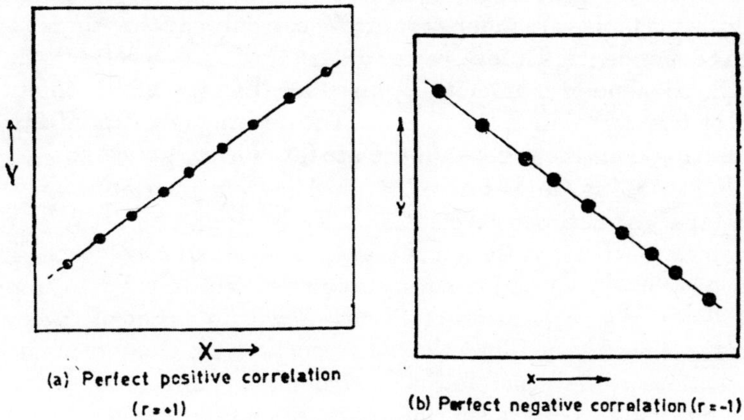

Fig. 7.2. Scatter diagrams of perfect correlation.

When the correlation coefficient is equal to + 1, all the points of the scatter diagram will fall on a definite straight line with positive slope (not necessarily 45°). Then the variables are said to be perfectly positively correlated. When the correlation coefficient is equal to – 1, all the points of the scatter diagram will fall on a straight line with a negative slope (not necessarily 45°). Then the variables are said to be perfectly negatively correlated. The cases of perfect positive and perfect negative correlations are depicted in Fig. 7.2 (*a*) and Fig. 7.2 (*b*) respectively.

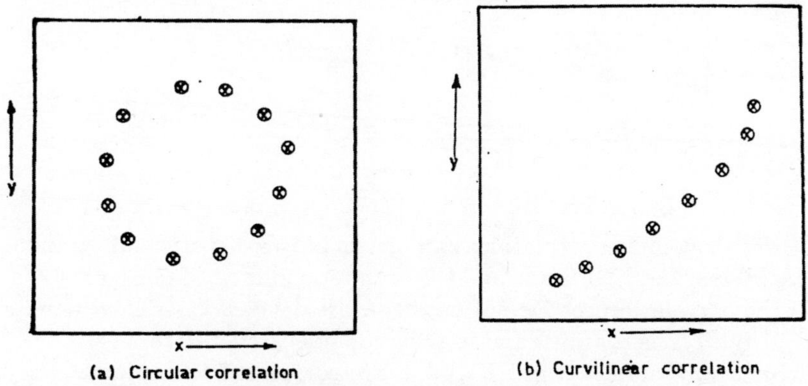

Fig. 7.3. Scatter diagrams of non-linear correlation.

It is to be noted here that the correlation coefficient as defined by Eq. (7.3) is a measure of only linear dependence. For example, consider the scatter diagram in Fig. 7.3 (*a*) and Fig. 7.3 (*b*). If we compute the correlation coefficient for these cases we may get, most likely, a value which is very near zero. But this does not mean at all that the variables are independent. We can only say that there is no linear dependence. Thus a scatter diagram will give a better insight into the dependence among the variables than the blind computation of the correlation coefficient. The techniques of analysis of measuring non-linear dependence are far more complicated.

Spurious Correlation. Even though two variables are uncorrelated, sometimes correlation may be apparent between the variables. Such a correlation is known as *spurious correlation*. The spurious correlation can arise due to clustering of data. For example, in Fig. 7.4 the correlation of Y with X within either of the data clusters is near zero. When the data from both the clusters are used to calculate a single correlation coefficient, this correlation is found to be quite high. This is spurious correlation.

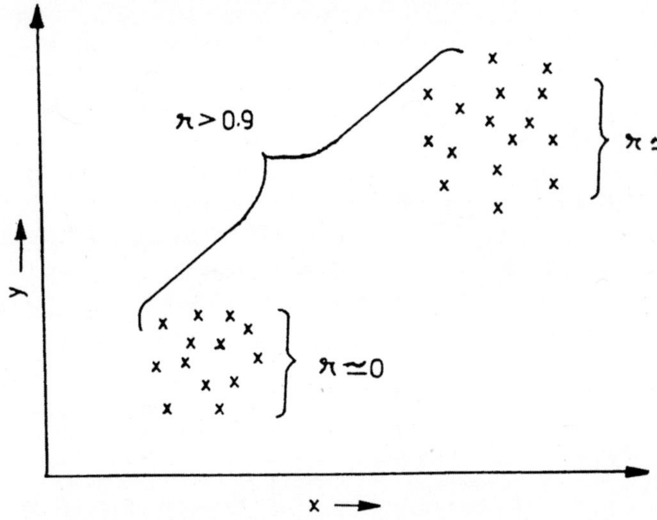

Fig. 7.4. Spurious correlation due to data clustering.

Spurious correlation can arise between ratios of random variables. Let X, Y and Z be three independent random variables. If the correlation coefficient between the derived variables such as X/Z and Y/Z is computed, it may be considerably different from zero which may indicate that X and Y are correlated when infact they are independent.

Coefficient of determination. The use of *coefficient of determination*, which is the square of coefficient of correlation, is perhaps a more convenient and useful way for interpreting the dependence between the variables. This is because the coefficient of determination gives the per cent of variation explained by one variable in the other.

$$\left.\begin{array}{l}\text{coefficient of}\\ \text{determination}\end{array}\right\} = r^2 = \frac{\text{Variance explained by } X \text{ in } Y}{\text{Variance in } Y} \quad ...(7.4)$$

Thus if $r = 0.9$, then $r^2 = 0.81$. Then it would mean 81% of the variance in variable Y is explained by the variable X due to linear dependence between them. Consider a case with $r = 0.6$ between one set of variables and the second case with $r = 0.3$ between another set of variables. We cannot say that the correlation in the second case is half as strong as the correlation in the first case. Because it implies that in the first case 36% of the variance is explained whereas the variance explained in the second case is only 9%.

Example 7.1. *The concurrent average yearly rainfall over a basin and the corresponding yearly runoff, both expressed in cm, for a period of 17 years are given below. Establish the dependence between yearly rainfall and yearly runoff by computing the coefficient of correlation between them.*

Year	*1*	*2*	*3*	*4*	*5*	*6*	*7*	*8*	*9*	*10*
rainfall	*113*	*128*	*127*	*104*	*108*	*115*	*167*	*154*	*99*	*119*
runoff	*74*	*104*	*96*	*61*	*59*	*82*	*109*	*102*	*57*	*78*
Year	*11*	*12*	*13*	*14*	*15*	*16*	*17*			
rainfall	*152*	*137*	*165*	*151*	*160*	*130*	*149*			
runoff	*109*	*96*	*124*	*103*	*134*	*87*	*106*			

Sol. Let the yearly rainfall be denoted by X and the yearly runoff be denoted by Y. The data and the necessary calculations are shown in Table 7.1.

Table 7.1. Computations for finding correlation coefficient of Example 7.1

i	*rainfall* x_i	*runoff* y_i	$(x_i - \bar{x})$	$(x_i - \bar{x})^2$	$(y_i - \bar{y})$	$(y_i - \bar{y})^2$	$(x_i - \bar{x})(y_i - \bar{y})$
1	113	74	− 21	441	− 19	361	339
2	128	104	− 6	36	11	121	− 66
3	127	96	− 7	49	3	9	− 21
4	104	61	− 30	900	− 32	1024	960
5	108	59	− 26	676	− 34	1156	884
6	115	82	− 19	361	− 11	121	209
7	167	109	33	1089	16	256	528
8	154	102	20	400	9	81	180
9	99	57	− 35	1225	− 36	1296	1260

10	119	78	− 15	225	− 15	225	225
11	152	109	18	324	16	256	288
12	137	96	3	9	3	9	9
13	165	124	31	961	31	961	961
14	151	103	17	289	10	100	170
15	160	134	26	676	41	1681	1066
16	130	87	− 4	16	− 6	36	24
17	149	106	15	225	13	169	195
Σ	2278	1581	0	7902	0	7862	7271

$$\bar{x} = \frac{1}{n} \Sigma x_i = \frac{2278}{17} = 134$$

$$\bar{y} = \frac{1}{n} \Sigma y_i = \frac{1581}{17} = 93$$

$$s_x = \sqrt{\frac{\Sigma (x_i - \bar{x})^2}{(n-1)}} = \sqrt{\frac{7902}{16}} = 22.22$$

$$s_y = \sqrt{\frac{\Sigma (y_i - \bar{y})^2}{(n-1)}} = \sqrt{\frac{7862}{16}} = 22.17$$

$$s_{xy} = \frac{1}{n} \Sigma(x_i - \bar{x})(y_i - \bar{y})$$

$$= \frac{7271}{17} = 427.7$$

$$r = \frac{s_{xy}}{s_x \cdot s_y} = \frac{427.7}{22.22 \times 22.17} = 0.868$$

$$r^2 = 0.754.$$

Since the correlation coefficient is very high there is every reason to believe that there is a strong linear dependence between X and Y. In other words if rainfall on the catchment is known in any year, the runoff of that year can be predicted with a fair amount of confidence.

Rank correlation coefficient. The *rank correlation coefficient* is also called *Spearman's rank correlation coefficient* and is usually denoted by r_s. For a given set of paired data $\{(x_i, y_i), i = 1, 2, ..., n\}$ it is obtained by ranking the x's among themselves and also the y's both from low to high or high to low and then using the formula

$$r_s = 1 - \frac{6 \sum_{i=1}^{n} d_i^2}{n(n^2 - 1)} \qquad ...(7.5)$$

where d_i is the difference between the ranks assigned to x_i and y_i. When there are ties in rank, the observations are assigned mean of the ranks which they jointly occupy. It can be shown that, when there are no ties in the rank, r_s is actually the correlation coefficient calculated for ranks. The rank correlation coefficient r_s is sometimes used instead of the correlation coefficient r mainly because it is easy to compute and is very close to r. In hydrology the rank correlation coefficient is very rarely used.

Example 7.2. *Compute the rank correlation coefficient for the rainfall runoff data of Example 7.1.*

Sol.

Table 7.2. Computations for determining rank correlation coefficient

i	x_i	rank of x_i	y_i	rank of y_i	difference in ranks d_i	d_i^2
(1)	(2)	(3)	(4)	(5)	(6)	(7)
1	113	14	74	14	0	0
2	128	10	104	6	4	16
3	127	11	96	9.5	1.5	2.25
4	104	16	61	15	1	1
5	108	15	59	16	– 1	1
6	115	13	82	12	1	1
7	167	1	109	3.5	– 2.5	6.25
8	154	4	102	8	– 4	16
9	99	17	57	17	0	0
10	119	12	78	13	– 1	1
11	152	5	109	3.5	1.5	2.25
12	137	8	96	9.5	– 1.5	2.25
13	165	2	124	2	0	0
14	151	6	103	7	– 1	1
15	160	3	134	1	2	4
16	130	9	87	11	– 2	4
17	149	7	106	5	2	4
						$\Sigma = 62$

The computations are shown in Table 7.2. The values in col. (2) and col. (4) are taken from Table 7.1. When ranks are assigned to the values of y, the value 96 occurs twice at 10th and 9th place and therefore both of them are given a rank of $(10 + 9)/2 = 9.5$.

Similarly among the values of y, 109 appears twice at 3rd and 4th positions and it is assigned a rank of 3.5.

$$r_s = 1 - \frac{6 \sum_{i=1}^{n} d_i^2}{n(n^2 - 1)} = 1 - \frac{6 \times 62}{17(17^2 - 1)} = 0.924.$$

7.2. REGRESSION

In many problems there are two or more variables that are inherently related and it may be necessary to explore the nature of this relationship. Regression analysis is a statistical technique for modelling and investigating the relation between two or more variables. Suppose that there is a *dependent variable* y that is related to k *independent variables* say $x_1, x_2, ..., x_k$. The relationship between these variables is characterised by a mathematical model called a *regression equation.* More precisely we say the regression of y on $x_1, x_2, ..., x_k$. This regression equation is fitted to a set of observed data. In some cases the investigator may know the exact form of the functional relationship between y and $x_1, x_2, ..., x_k$, say $y = \phi\,(x_1, x_2, ..., x_k)$. However, in most cases, the true functional relationship is unknown and the investigator will choose an appropriate function to approximate ϕ. A polynomial model is usually employed as the appropriate function.

Regression analysis reveals the average relationship between the variables. It refers to the method by which estimates are made of the values of a variable from the knowledge of the values of one or more other variables. We must note here the following essential difference between correlation and regression. Correlation merely ascertains the degree of linear dependence between the two variables. It does not say that one variable is the cause (independent variable) and the other variable is the effect (dependent variable). However, in the regression analysis one variable is taken as dependent and the other as independent thus making it possible to study the cause and effect relationship. It is for the investigator to choose, based on his physical understanding of the situation under consideration, which of the variables are independent and which other is independent. The presence of correlation between two variables doe not necessarily imply a cause and effect relationship between them whereas the existence of cause and effect relationship between two variables does imply correlation between them. If one computes the correlation coefficient using the data of simultaneous observations on the number of hail storms experienced in Hyderabad of

South India in a year and the number of earth-quakes recorded in Japan in the same year he may get a significant value for the same. But as we know there is no cause and effect relationship between these two variables and therefore they are totally uncorrelated.

Simple linear regression. Here we discuss the case where only a single independent variable X is of interest. For example, we may like to find the regression equation between yearly rainfall on a basin (taken as an independent variable) and the corresponding yearly runoff from the same basin (considered as dependent variable). Linear regression equation can be fitted between two variables based on the observed data if it is felt that the relation between these two variables is linear. Let X and Y be the independent and dependent variables respectively. Let (x_1, y_1), (x_2, y_2), ..., (x_n, y_n) be the pairs of concurrent observations taken on X and Y. We assume the following relation between these variables.

$$Y = a + bX \qquad ...(7.6)$$

where a and b are the constants to be obtained from the regression analysis. The predicted value $\hat{y}_i$ from this equation corresponding to any observed value x_i is thus given by $\hat{y}_i = a + bx_i$. If there is perfect correlation between X and Y, the best combination of a and b will be such that $\hat{y}_i = y_i$. When there is no perfect dependence between X and Y, then however best the values of a and b may be the prediction cannot be exact and there would be always a deviation between the predicted and observed values of Y. Fig. 7.5 shows these deviations.

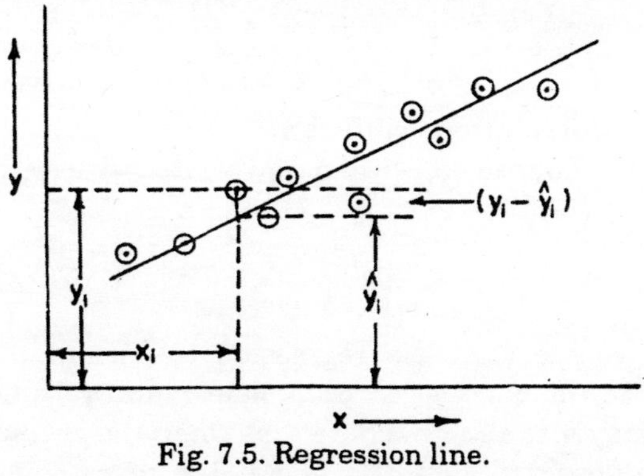

Fig. 7.5. Regression line.

Since $\hat{y}_i = a + bx_i$, the deviation $(y_i - \hat{y}_i)$ is given by $(y_i - a - bx_i)$. Obviously the best regression line is the one which makes these deviations as small as possible. The usual criterion that is adopted to obtain the best values of a and b is to minimise the sum of the squared deviations over all observations. That is minimise $S = \sum_{i=1}^{n} (y_i - a - bx_i)^2$. The values of a and b which minimise S are then obtained from the equations

$$\frac{\partial S}{\partial a} = 2\Sigma\,(y_i - a - bx_i)\,(-1) = 0$$

$$\frac{\partial S}{\partial b} = 2\,\Sigma\,(y_i - a - bx_i)\,(-x_i) = 0$$

These equations are reduced to

$$\Sigma\,(y_i - a - bx_i) = 0 \qquad \text{...(7.7)}$$

$$\Sigma\,x_i\,(y_i - a - bx_i) = 0 \qquad \text{...(7.8)}$$

Eqs. (7.7) and (7.8) are called the *normal equations* of linear regression.

Eq. (7.7) gives the solution to a as

$$a = \bar{y} - b\,\bar{x} \qquad \text{...(7.9)}$$

Substitution of Eq. (7.9) in Eq. (7.6) yields a very convenient form of regression line

$$(y - \bar{y}) = b(x - \bar{x}) \qquad \text{...(7.10)}$$

Eq. (7.10) indicates that the regression line passes through the point of averages $(\bar{x}, \bar{y})$. The solution for b is obtained from Eqs. (7.7) and (7.8) as

$$b = \frac{\frac{1}{n}\Sigma\,x_i\,y_i - \bar{x}\,\bar{y}}{\frac{1}{n}\Sigma\,x_i^{\,2} - \bar{x}_2} = \frac{s_{xy}}{s_x^{\,2}}$$

But the correlation coefficient r between X and Y is given by $r = s_{xy}/(s_x . s_y)$. Therefore, we can now write, with s_{xy} replaced by $r\,.\,s_x\,.\,s_y$,

$$b = \frac{r\,.\,s_x\,.\,s_y}{s_x^{\,2}} = r\,\frac{s_y}{s_x} \qquad \text{...(7.11)}$$

Thus we see that the regression coefficient b is directly proportional to the correlation coefficient r and retains the same sign as r. It may also be noted here that when the regression line is fitted for the observed data of X and Y, the regression constant b denotes the slope of the regression line and other constant a denotes the intercept on y-axis. Further one should note that the correlation coefficient r does not indicate the slope.

If the analysis is carried out taking Y as the independent variable and X as the dependent variable, we get a similar result

$$(x - \bar{x}) = b' (y - \bar{y}) \qquad \text{...(7.12)}$$

$$b' = r \frac{s_x}{s_y} \qquad \text{...(7.13)}$$

Since the regression line cannot make perfect prediction, we need to define a measure which would indicate how precise the prediction is. We define the residuals as $e_i = (y_i - \hat{y}_i)$, $i = 1, 2, \ldots\ldots, n$. The square root of the sum of the squared residues is known as the *standard error of estimate*. Obviously, the lesser the standard error of estimate the better is the prediction. It can be shown that the standard error of estimate is given by

$$s_e = \sqrt{(1 - r^2)\, s_y^{\,2}} \qquad \text{...(7.14)}$$

As pointed out earlier, r^2 is known as the coefficient of determination. From Eq. (7.14) we notice that higher the coefficient of determination lesser is the standard error of estimate. This explains the significance of the coefficient of determination.

Example 7.3. *Obtain the regression line to predict the annual runoff based on the known values of annual rainfall from the data of Example 7.1.*

Sol. Let Y and X denote the annual runoff and the annual rainfall of the basin respectively. Then the required regression equation is

$$(y - \bar{y}) = b\,(x - \bar{x}).$$

From the computations of Example 7.1, we have

$\bar{x} = 134$, $s_x = 22.222$, $\bar{y} = 93$, $s_y = 22.17$, and $r = 0.868$.

Now $$b = \frac{r.s_y}{s_x} = 0.868 \frac{22.17}{22.22} = 0.866.$$

Therefore $(y - 93) = 0.866\,(x - 134)$

or $$y = 93 + 0.866x - 116$$

$$y = 0.866x - 23.0$$

is the required regression line with a standard error of estimate of

$$s_e = \sqrt{(1 - 0.868^2)\, 22.17^2} = 11.0 \text{ cm}.$$

Transformations. Sometimes we find that the linear regression model of the form $y = ax + b$ is inappropriate because the true regression function is non-linear. This is sometimes visible from the scatter diagram, and sometimes we know in advance that the relation is non-linear because of earlier experience or because of the underlying theory. For example, one would not be justified in fitting a linear regression equation between the stage in the river and the discharge carried by the river. In some of such situations a non-linear function can be expressed as a straight line by using a suitable transformation. Such non-linear models are called *intrinsically linear models.*

If we prepare a scatter diagram between the logarithms of the variables instead of the variables themselves and if the resulting plot appears to fit a near straight line, we can choose the equation of the form $y = a\,x^b$. This function is intrinsically linear, since it can be transformed to a straight line by a logarithmic transformation $\ln y = \ln a + b \ln x$.

If the logarithm of the dependent variable is plotted against the independent variable and if the resulting scatter diagram indicates a straight line, the relation to be used is $y = a\,e^{bx}$. Because this intrinsically linear function can be transformed to a straight line as $\ln y = \ln a + bx$.

Another intrinsically linear function is $y = a + b/x$. By using a reciprocal transformation $z = 1/x$, the model is linearised to $y = a + bz$.

Sometimes the logarithmic and reciprocal transformations can be employed jointly to linearise a function. For example, consider the function $y = 1/(\exp(a + bx)$. Letting $y^* = 1/y$, we have the linearised form $\ln y^* = a + bx$. Several other non-linear models that are intrinsically linear may be found.

Since the relations are linear after transformation, the regression analysis can now be performed on the transformed variables to obtain the required regression constants a and b.

Polynomial fitting. Curvilinear data which do not indicate a linear scatter diagram even under any of the transformations discussed above can be fitted by a polynomial of the form

$$y = a_0 + a_1 x + a_2 x^2 + \ldots\ldots + a_p x^p \qquad \ldots(7.15)$$

where $a_0, a_1, \ldots\ldots, a_p$ are the regression coefficients to be determined and p is called the degree of the polynomial. Equating the partial derivatives of the sum of the squared deviations S with respect to $a_0, a_1, \ldots\ldots, a_p$ to zero as in the case of simple linear regression, we get the following equations.

$$\left.\begin{aligned}
\Sigma y_i &= a_0 n + a_1 T^1 + a_2 T^2 + \ldots\ldots\ldots\ldots + a_p T^p \\
\Sigma y_i x_i &= a_0 T^1 + a_1 T^2 + a_2 T^3 + \ldots\ldots\ldots + a_p T^{p+1} \\
\Sigma y_i x_i^2 &= a_0 T^2 + a_1 T^3 + a_2 T^4 + \ldots\ldots\ldots\ldots + a_p T^{p+2} \\
&\vdots \\
\Sigma y_i x_i^p &= a_0 T^p + a_1 T^{p+1} + a_2 T^{p+2} + \ldots\ldots + a_p T^{2p}
\end{aligned}\right\} \qquad \ldots(7.16)$$

$$\text{where } T^k = \sum_{i=1}^{n} x_i^{\,k} \qquad ...(7.17)$$

and n is the number of observed pairs of data. Equations in (7.16) are called the *normal equations* of polynomial regression. There will be $(p + 1)$ simultaneous equations in (7.16) the solution of which yields the values of $a_0, a_1, a_2, ..., a_p$.

The degree of the polynomial p can be so chosen that the polynomial fit is as satisfactory as required and at the same time it does not become unwiedly for the purpose of predicting Y. It may be noted here that as we increase the degree of polynomial, the number of summations to be evaluated T^k and the number of equations required for the solution of the regression constants also increase making the computations more and more tedious. Therefore we have to resort to some sort of compromise in choosing the degree of the polynomial. The procedure often adopted is to try with $p = 2$ onwards. After fitting the second degree polynomial, the correlation coefficient between $\hat{y}_i$ and y_i is computed. This is called the *coefficient of multiple correlation*. The nearer the value of multiple correlation coefficient to 1, the better is the prediction by polynomial fitting. Every time the degree of the polynomial is increased by 1 and the coefficient of multiple correlation is determined. The procedure is terminated when the value of the multiple correlation coefficient is large enough, say more than 0.9, or when the increase in the value of multiple correlation in two successive iterations is insignificant. This fixes the desired degree of polynomial.

One should be cautious in using the regression equations for extrapolation. If the actual relationship is curvilinear with a large radius of curvature, the observed data forming a small segment of this curve may indicate a linear relationship and predictions very much outside the observed range may prove grossly erroneous.

7.3. MULTIPLE REGRESSION

A regression model that involves more than one independent variable is called a *multiple regression model*. For example, the peak runoff from a catchment may be related to the catchment area, slope of the catchment, rainfall rate etc. A multiple regression equation describing the relationship between the dependent variable Y and the independent variables $X_1, X_2, ..., X_p$ may be of the form

$$y = b_0 + b_1x_1 + b_2x_2 + ... + b_px_p \qquad ...(7.18)$$

This is called a *multiple linear regression equation*. The term *linear* is used because Eq. (7.18) is a linear function of the unknown parameters $b_0, b_1, ..., b_p$. Therefore, the regression equation

$$y = b_0 + b_2x_2 + b_3x_3{}^2 + b_4 (\ln x_4)$$

is also linear in parameters b_0, b_2, b_3 and b_4. But the equations

$$y = b_1 + b_2 x^{b_3}$$
$$y = b_1 + b_2x_2 + b_3 \exp (b_4 x_3)$$
$$y = b_1 + b_2x_2 + b_2{}^2 x_2$$

are not linear in the parameters. The parameters $b_1, b_2, ..., b_p$ in Eq. (7.18) are often called *partial regression coefficients*, because they describe the partial effect of one independent variable when the other independent variables in the equation are held constant. Multiple linear regression models are often used as approximating functions. That is, the true functional relationship between y and $x_1, x_2, ..., x_p$ is unknown, but over certain range of the independent variables the linear regression model is a good enough approximation.

To obtain the values of the regression coefficients in practice, n observations would be available on Y along with the corresponding n observations on each of the p independent variables. Thus n equations like Eq. (7.18) can be written one for each observation as given under.

$$\left.\begin{array}{l} y_1 = b_0 + b_1x_{11} + b_2 x_{21} + ... + b_px_{p_1} \\ y_2 = b_0 + b_1x_{12} + b_2x_{22} + ... + b_px_{p_2} \\ y_i = b_0 + b_1x_{1i} + b_2x_{2i} + ... + b_px_{p_i} \\ \vdots \quad \vdots \quad \vdots \quad \vdots \quad \vdots \\ y_n = b_0 + b_1x_{1n} + b_2x_{2n} + ... + b_px_{p_n} \end{array}\right\} \qquad ...(7.19)$$

where y_i is the ith observation on Y and x_{ji} is the corresponding ith observation on jth independent variable x_j. Then we will essentially be solving n simultaneous equations for the $(p + 1)$ unknown parameters. Thus n must be equal to or greater than $(p + 1)$. In practice n should be at least 3 or 4 times as large as $(p + 1)$. Equation (7.19) can be written in a concise form as

$$y_i = b_0 + \sum_{j=1}^{p} b_j x_{ji} \qquad ...(7.20)$$

$$i = 1, 2, ..., n.$$

If the method of least squares is used to estimate the regression coefficients of Eq. (7.20), we obtain the following normal equations in which all the summations run from $i = 1$ to $i = n$.

$$\left.\begin{array}{l}\Sigma y_i = b_0 n + b_1 \Sigma x_{1i} + b_2 \Sigma x_{2i} + \ldots\ldots + b_p \Sigma x_{pi} \\ \Sigma y_i x_{1i} = b_0 \Sigma x_{1i} + b_1 \Sigma x_{1i}^2 + b_2 \Sigma x_{2i} x_{1i} + \ldots\ldots + b_p \Sigma x_{p_i} x_{1i} \\ \vdots \\ \Sigma y_i x_{pi} = b_0 \Sigma x_{pi} + b_1 \Sigma x_{1i} x_{pi} + b_2 \Sigma x_{2i} x_{pi} + \ldots\ldots + b_p \Sigma x_{pi}^2 \end{array}\right\} \quad \ldots(7.21)$$

There are $(p + 1)$ normal equation which when solved will provide the solution for the unknown regression coefficient b_0, b_1, b_2, ..., b_p.

It is simple to solve the normal equations if they are expressed in matrix notation. Let the matrices be denoted by the corresponding bold face characters. Then Eq. (7.20) can be written in the matrix notation as

$$\mathbf{y} = \mathbf{xb} \quad \ldots(7.22)$$

where $\mathbf{y} = \begin{bmatrix} y_1 \\ y_2 \\ \vdots \\ y_n \end{bmatrix}$, $\mathbf{x} = \begin{bmatrix} 1 & x_{11} & x_{21} \ldots\ldots & x_{p1} \\ 1 & x_{12} & x_{22} \ldots\ldots & x_{p2} \\ \vdots & \vdots & \vdots & \vdots \\ 1 & x_{1n} & x_{2n} \ldots\ldots & x_{pn} \end{bmatrix}$, $\mathbf{b} = \begin{bmatrix} b_0 \\ b_1 \\ b_2 \\ \vdots \\ b_p \end{bmatrix}$

In general, $\mathbf{y}$ is a $(n \times 1)$ column matrix of the observations on Y, $\mathbf{x}$ is a $(n \times k))$ matrix with all the elements in the first column as 1s and the remaining $(k - 1)$ columns being the observations on the p independent variables X_1, X_2, ..., X_p. Thus $k = p + 1$. And $\mathbf{b}$ is a $(p \times 1)$ column matrix of the regression coefficients. In terms of these matrices, it can be shown that the solution of the regression coefficients is given by

$$\mathbf{b} = (\mathbf{x}'\,\mathbf{x})^{-1}\,\mathbf{x}'\mathbf{y} \quad \ldots(7.23)$$

where $\mathbf{x}'$ is the transpose of $\mathbf{x}$ and $(\mathbf{x}'\,\mathbf{x})^{-1}$ is the inverse of the matrix $(\mathbf{x}'\,\mathbf{x})$.

In some cases, even if the relationship between dependent variable and the independent variables cannot be assumed to be of the form of multiple linear regression equation, it can be brought to a linear form by a simple logarithmic transformation. For example, the relationship

$$y = b_0\, x_1^{\,b_1}\, x_2^{\,b_2} \ldots\ldots x_p^{\,b_p}$$

may be brought to a linear form after taking logarithms on both sides.

$$\ln y = \ln b_0 + b_1 \ln x_1 + b_2 \ln x_2 + \ldots\ldots + b_p \ln x_p$$

Now the multiple linear regression can be carried out on the transformed variables. Similarly consider the cubic polynomial equation in one independent variable

$$y = b_0 + b_1 x + b_2 x^2 + b_3 x^3.$$

If we let $x_1 = x$, $x_2 = x^2$ and $x_3 = x^3$, then it can be written as

$$y = b_0 + b_1 x_1 + b_2 x_2 + b_3 x_3$$

which is a multiple linear regression model. Models which include interaction effect may also be analysed by multiple linear regression method. For example, suppose the model is

$$y = b_0 + b_1 x_1 + b_2 x_2 + b_{12} x_1 x_2$$

If we let $x_3 = x_1 x_2$ and $b_3 = b_{12}$, then it can be written as

$$y = b_0 + b_1 x_1 + b_2 x_2 + b_3 x_3$$

which is a multiple linear regression model.

Measures of Adequacy. Verifying the adequacy of the model is an important part of the multiple regression model building. A number of techniques can be used to measure the adequacy of the multiple regression model. We will now describe one such measure which is commonly used.

After the regression coefficients are obtained, for every observed $(x_{1i}, x_{2i}, \ldots\ldots, x_{pi})$ the value of $\hat{y}_i$ can be computed using Eq. (7.20) for $i = 1, 2, \ldots\ldots, n$. Obviously $\hat{y}_i$ cannot be exactly equal to observed y_i for each i. The correlation coefficient between $\hat{y}_i$ and y_i is called the *coefficient of multiple correlation* and is denoted by R_p where p is the number of independent variables used in the multiple linear regression. The square of the coefficient of multiple correlation, R_p^2, is called the *coefficient of multiple determination*. We have that $0 \le R_p^2 \le 1$. A large value of R_p^2 indicates that the regression model is a good one.

Selection of variables. An important problem in many applications of regression analysis is the selection of the set of independent variables to be used in the model. Sometimes the previous experience or the underlying theoretical considerations can help the analyst to choose a proper set of independent variables. Usually, however, the problem of selecting an appropriate set of independent variables is more complex. The general situation is, we have a set of independent variables that quite likely includes all of the important variables and we are not sure if all of these variables are necessary to adequately describe the dependent variable. In such situations we have to screen all the variables to obtain a regression model that contains the best subset of independent variables. There is no technique which always produces a good solution to the variable selection problem. Most of the available procedures are search techniques requiring interaction and judgement on the part of analyst. The commonly used variable selection techniques are *all possible regressions* and *stepwise regression*.

In the method of all possible regressions, the analyst fits all of the regression equations involving one variable, all regression equations involving two variables and so on. If there are k independent variables to be considered then there will be 2^k equations to be developed. For example, if there are 3 independent variables then all possible regression equations to be developed are between y and x_1 ; y and x_2 ; y and x_3 ; y and x_1, x_2 ; y and x_2, x_3 ; y and x_3, x_1 ; y and x_1, x_2, x_3. Including the trivial equation with only an intercept there will be 2^3, that is, 8 equations. Then these equations are evaluated according to some suitable criterion to select the "best" regression equation. In this procedure, the number of equations to be examined increases rapidly as the number of independent variables increases. For example, if $k = 10$, there are $2^{10} = 1024$ possible regression equations. Therefore all possible regressions method requires considerable computational effort.

Stepwise regression is possibly the most widely used variable selection method. This procedure consists of building the regression equation by adding or deleting one variable at a time at each step. The first variable added is the one with the highest simple correlation with the dependent variable. The second variable added is the one explaining largest variation in the dependent variable that remains unexplained by the first variable added. At this point, the first variable is tested for significance and retained or discarded depending on the result of the test. The third variable added is the one that explains the largest portion of the variation that is not explained by the variables already in the equation. The variables in the equation are then tested for significance. This procedure is continued until all of the variables not in the equation are found to be insignificant and all of the variables in the equation are significant. This is a very good procedure to use but care must be taken to see that the resulting equation is rational.

EXERCISES

7.1. The yearly rainfall and the yearly runoff of a catchment for a period of 16 years is given below. Determine the correlation coefficient between yearly rainfall and yearly runoff. Also compute the rank correlation coefficient between them.

Year	*Rainfall (mm)*	*Runoff (mm)*
1976	1076	337
1977	850	84
1978	1210	385
1979	1276	394
1980	1100	361
1981	1336	538
1982	789	196
1983	1270	448
1984	1196	582
1985	1196	480
1986	1037	326
1987	948	294
1988	944	385
1989	1026	264
1990	1153	458
1991	1042	413

7.2. Calculate the regression equation for the data of Exercise 7.1 considering rainfall as the independent variable.

7.3. Calculate the regression equation for the data of Exercise 7.1 considering runoff as the independent variable. Does this equation agree with the one obtained in Exercise 7.2 ? Comment.

7.4. Show that the correlation coefficient between y and $\hat{y}$ in the simple linear regression is same as the correlation between y and x.

7.5. Derive the normal equation for b to fit a linear regression equation assuming that it has to pass through the origin.

7.6. Derive the normal equations which minimise the sum of the squared deviations for the regression equation $y = a\,x^b$.

7.7. Explain what do you understand by spurious correlation, coefficient of determination, coefficient of multiple correlation, and coefficient of multiple determination.

7.8. Develop a regression equation of the form $Q = a\,G^b$ between the discharge in the river Q and the corresponding stage G using the data given below.

Q in m^3/s	G in m	Q in $m^3/3$	G in m
1688	4.72	640	2.69
1470	4.30	578	2.53
1589	4.54	1093	3.57
1481	4.31	1053	3.47
1535	4.30	1028	3.44
1419	4.11	875	3.14
1625	4.43	968	3.32
1317	3.95	937	3.26
1232	3.84	818	3.07
1178	3.72	731	2.89
1226	3.80	640	2.68
1249	3.83	515	2.37
1267	3.86		

7.9. Define the coefficient of correlation between two random variables X and Y and explain its significance. If two random variables are uncorrelated, that is if the correlation coefficient between the variables is zero, are they necessarily independent ? Why or why not ?

7.10. Derive the normal equations for simple linear regression and hence the equations for the two regression coefficients. Show that the regression equation passes through the point of average, that is $(\bar{x}, \bar{y})$.

8

Stochastic Processes

8.1. INTRODUCTION

A variable associated with a random phenomenon, or a chance phenomenon as it is often called, is a random variable whose behaviour is completely described by its probability distribution function. In such a description an assumption is implied that the value that a random variable can take on in any observation is independent of all the past observations nor has it any bearing on the future observations. The same probability distribution can be used to predict its outcome at any time. In such circumstances the phenomenon may be called a *pure random process*. In other words, in a pure random process the probability distribution does not vary with time, or we say the probability law is time invariant. There are many processes whose outcome cannot be predicted uniquely and hence they come under the general category of probabilistic phenomena, but their outcomes are not truly independent in occurrence. We cannot employ the same probability distribution as it changes with time. For example, the magnitude of the monthly run off may depend, to some degree, on the magnitude of the run off in the previous months and therefore the probability that the monthly run off in the stream exceeds a certain magnitude in the month of october is not the same as that in April. In other words, there is a some kind of dependence among the occurrences of the variable at different times and the probability distribution is not time invariant. Such phenomena may be broadly categorised as a *stochastic process*. A more precise definition and classification of stochastic processes now follows.

8.2. CLASSIFICATION OF STOCHASTIC PROCESSES

A phenomenon which changes continuously with time may be called a *process*. We use a variable $X(t)$ to quantify the state of the process at time t. Thus, the variable X changes as a function of time. In some other situations it may vary as a function of some other physical parameter, say, space. For example, the thickness of a wire may vary in some random manner with distance along the length of the wire.

If the value of the variable X can be predicted uniquely for any time t through a deterministic function of time then the process is called a *deterministic process*. For example, in the majority of electrical engineering calculations the supply voltage is represented by

$$X(t) = a \cos (2\pi f_0 t + \phi) \qquad ...(8.1)$$

where f_0 is the supply frequency, a is the voltage amplitude and ϕ is the phase angle with respect to time origin.

On the other hand if the value of the variable cannot be predicted uniquely the process is called a *stochastic process* or more commonly a *random process*. The behaviour of a stochastic process can be described by a set of random variables $\{X(t)\}$, where the time variable t can take any value from $-\infty$ to ∞. Thus the statistical properties of a stochastic process are described by associating a joint probability distribution with any set of times $t_1, t_2, \ldots\ldots, t_n$. In other words, *the ordered set of random variables* $\{X(t_1), X(t_2), \ldots\ldots, X(t_n)\}$ *and its associated joint probability distribution is called a stochastic process.*

Stochastic processes are usually classified as to whether the random variables associated with the process are discrete or continuous and whether the indexing variable t is discrete or continuous. If the random variables are discrete the process is usually referred to as a *discrete state stochastic process* the examples of which may be given as Markov chains, Poisson process, renewal process etc. If the random variables can take any real number then the process is referred to as *continuous state stochastic process.* If a process exists continuously in time then it is called a *continuous time stochastic process*. For discrete time processes there are two possibilities. The underlying process may exist in continuous time, but the observations are made only at discrete time points, or they are obtained by averaging the values over the time intervals or by any rule of selecting particular value over discrete time intervals. Or the process may be truly discrete yielding observations only at discrete points in time. In all these cases the process is said to be a *discrete time stochastic process.*

In all, we have the following four types of stochastic processes as combinations of the above two classifications :

(*i*) *Continuous state-continuous time stochastic process*

$\{X(t)\}$ with $-\infty < X(t) < \infty$; and $-\infty < t < \infty$

(*ii*) *Continuous state discrete time stochastic process*

$\{X(t)\}$ with $-\infty < X(t) < \infty$ and $t = 0, \pm \Delta t, \pm 2\Delta t, \pm 3\Delta t, ...$

where Δt is the sampling time interval.

(*iii*) *Discrete state continuous time stochastic process*

$\{X(t)\}$ with $X(t) = x_1$ or x_2 or x_3, ..., or x_n and $-\infty < t < \infty$

where $x_1, x_2, ..., x_n$ are n possible values which the random variables can take on and n is finite or countably infinite.

(*iv*) *Discrete state-discrete time stochastic process*

$\{X(t)\}$ with $X(t) = x_1$ or x_2 or x_3 ...... or x_n

and $t = 0, \pm \Delta t, \pm 2\Delta t, \pm 3\Delta t, ...$

Majority of hydrologic processes are continuous state-continuous time stochastic processes. Depth to ground water table at a given location, the elevation of water surface in a lake, the run off discharge of a perennial river, the water temperature etc., may be given as examples of continuous state continuous time stochastic process since the random variables quantifying them can take on any value in the range of fluctuations at any time. For practical purposes, however, they are approximated as continuous state discrete time stochastic processes.

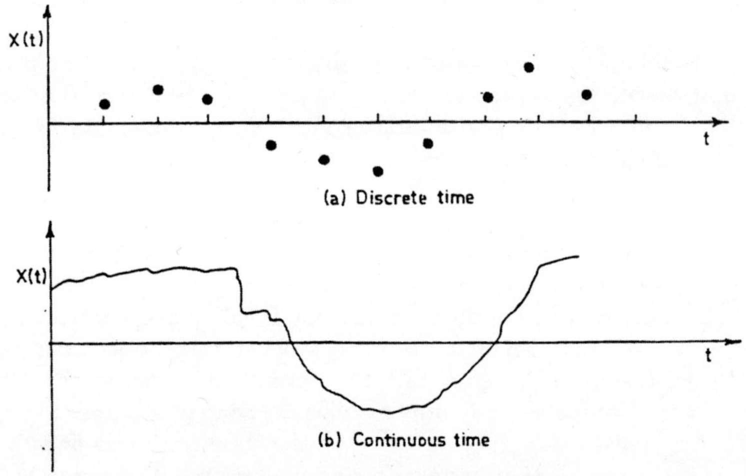

Fig. 8.1. Examples of time series.

If a stochastic process is observed, or partially observed the set of observations is called a *time series* or a *realisation* or a *sample function*. The difference, then, between a stochastic process and a time series is the same as the difference between a random variable and an observation of the random variable. With regard to notation, as in the case of random variables, an observed value of $X(t)$ will be denoted by $x(t)$ in the continuous time case and when the time is discrete these will be denoted by X_t and x_t respectively. A time series or a set of observations is a sequence of numbers or a curve which can be plotted as shown in Fig. 8.1.

8.3. STATIONARY PROCESS

An important characteristic that a stochastic process may possess is *stationarity*. A stochastic process is said to be *strictly stationary* if for any n and any time points, t_i, the joint probability distribution of $X(t_1)$, $X(t_2)$, ..., $X(t_n)$ is identical with the joint probability distribution of $X(t_1 + \tau)$, $X(t_2 + \tau)$, ..., $X(t_n + \tau)$. In the continuous time case, τ is any real number representing the translation in time. In the discrete time case τ is any integral multiple of the sampling time interval. τ is also known as the *lag time*. In other words, this definition says that the probability distribution of the process at any time point is identical with the distribution at any other time point (where $n = 1$). The joint distribution at any pair of time points is identical with the joint distribution at any other pair of time points which are separated by the same time interval and so on and so forth. Absolute time plays no role in these joint distributions. Only the translation time or lag has influence on these joint distributions. In a *non-stationary* stochastic process, on the other hand, statistical properties differ from one segment to another and therefore are time dependent.

In many applications, complete invariance of the probability distribution with respect to translation in time is not necessary. A much weaker condition is to specify invariance with respect to first two moments only. This, of course, requires that second moment exists which is not necessary in the definition of a strictly stationary process. Accordingly a stochastic process is said to be *weakly stationary* if it is stationary only in mean and covariance. Thus, in a weakly stationary stochastic process we have that

$$E[X(t_1)] = E[X(t_2)] = ... = \mu, \text{ a constant}$$

$$E[X^2(t)] < \infty$$

and $\quad E[\{X(t_1) - \mu\}\{X(t_2) - \mu\}] = E[\{X(t_1 + \tau) - \mu\}\{X(t_2 + \tau) - \mu\}]$

for all t_1 and t_2 and any τ.

Since the normal distribution is completely determined by its first two moments, in the case of a Gaussian process weak stationarity implies strict stationarity. Also, strict stationarity plus the existence of second moment implies weak stationarity. Strict stationarity is also referred to as *strong stationarity*. Similarly weak stationarity is also referred to as *covariance stationarity* or *stationarity in wide sense*. It is interesting to note here that a process can be strictly stationary and yet it is not wide sense stationary if the moments $E[X(t)]$ and $E[X(t) \cdot X(t+\tau)]$ do not exist. On the other hand, a process can be wide-sense stationary and yet it is not strictly stationary since a higher-order probability distribution function can be dependent on time.

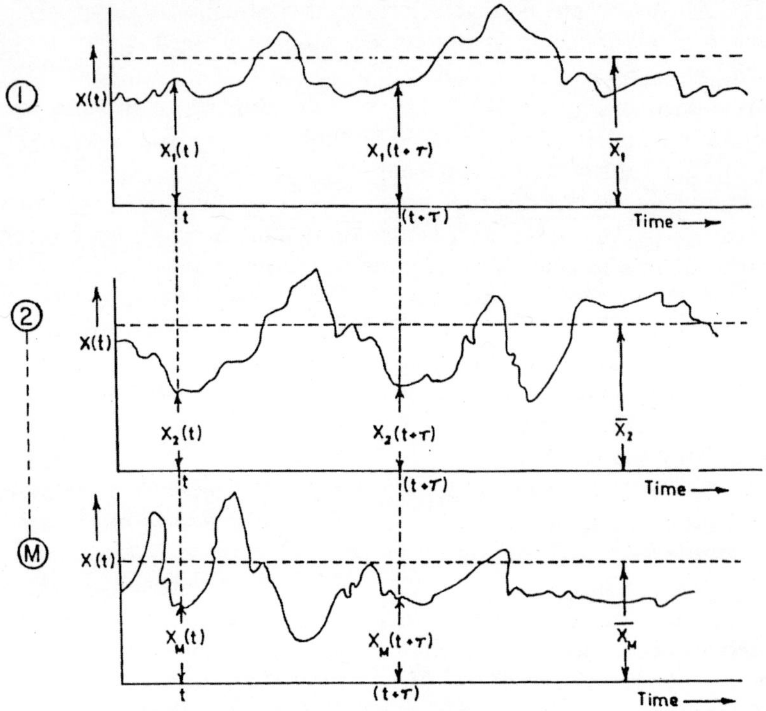

Fig. 8.2. Ensemble of continuous time series.

Stationary processes can be subdivided into *ergodic* and *non-ergodic* processes. In a stationary stochastic process ergodicity is the property by which each realisation of the process is a complete and independent representative of possible realisation of the process. Such realisations constitute an *ensemble* which is theoretically an infinite number of possible sets of observations. Theoretical consideration of ensemble of time series rather than a single realisation is analogous to the population and sample in statistics.

To explain further the concept of ergodicity and stationarity, consider an ensemble consisting of collection of m realisations of a stochastic process each of length T as shown in Fig. 8.2. Here the series $x(t)$ is assumed to be continuous in the range $0 \leq t \leq T$, though the same definitions can be applied equally well to an ensemble of discrete time series. The mean of $x(t)$ at the time position t from the series origin is given by

$$\bar{x}(t) = \frac{1}{m} \sum_{i=1}^{m} x_i(t) \quad ...(8.2)$$

and the mean of the series at the time position $(t + \tau)$ is given by

$$\bar{x}(t + \tau) = \frac{1}{m} \sum_{i=1}^{m} x_i (t + \tau) \quad ...(8.3)$$

The covariance of $x(t)$ and $x(t + \tau)$ is given by

$$\text{cov}\,[x(t), x(t + \tau)] = \frac{1}{m} \sum_{i=1}^{m} [\{x_i(t) - \bar{x}(t)\} \{x_i (t + \tau) - \bar{x}(t + \tau)\}] \quad ...(8.4)$$

If the means of $x(t)$ and $x(t + \tau)$ denoted by $\bar{x}(t)$ and $\bar{x}(t + \tau)$ and given by Eqs. (8.2) and (8.3) converge with probability 1 to the population mean μ as $m \to \infty$, that is, if $P\,[\bar{x}(t) = \mu] = 1$ as $m \to \infty$ regardless of the position of t from 0 to T, the process is said to be stationary in the mean or first order stationary. If the covariance given by Eq. (8.4) is independent of the position of t, but dependent only on the lag time, τ, and for a given lag τ if it approaches a constant defined by $\text{cov}\,[x(t), x(t + \tau)] = f(\tau) = \rho_\tau \sigma_x^2$ as $m \to \infty$, where σ_x^2 is the population variance of $X(t)$ and ρ_τ is the population correlation coefficient between $X(t)$ and $X(t + \tau)$, then the process is said to be stationary in covariance. Incidentally ρ_τ is called the *autocorrelation coefficient* of lag τ since it is the correlation coefficient between the same random variable situated at two different time positions separated by τ. If the process is stationary both in mean and covariance, then it is said to be second order stationary. Strict stationarity requires that all higher moments of $x(t)$, $x(t + \tau_1)$, $x(t + \tau_2)$ ...... are independent of t, but dependent on τ_1, τ_2, ... and they all converge with probability 1 to the corresponding higher population moments as $m \to \infty$.

Parameters of each particular series of the ensemble designated as the ith series can be estimated as

$$\bar{x}_i = \frac{1}{T} \int_0^T x_i (t)\, dt \quad .(8.5)$$

$$\text{cov}\,[x_i(t), x_i(t + \tau)] = \frac{1}{(T - \tau)} \int_0^{T - \tau} \left\{ x_i (t) - \bar{x}_i \right\} \left\{ x_i (t + \tau) - \bar{x}_i \right\} dt \quad ...(8.6)$$

For any series i, where $i = 1, 2, ..., m$, if $\bar{x}_i \rightarrow \mu$ with probability 1 as $T \rightarrow \infty$, and if Cov $[x_i (t)\, x_i (t + \tau)] = f(\tau) = \rho_\tau \sigma_x^2$ for any τ, and if this is valid for other parameters computed from higher order moments, then the process is said to be *ergodic*. Thus, for ergodic processes the time averaged mean value and autocorrelation coefficient as well as other time-averaged properties are equal to the corresponding ensemble averaged values. It is to be noted here that only a stationary stochastic process can be ergodic. Ergodic processes are clearly an important class of stochastic processes since all the properties of an ergodic process can be determined by performing time averages over a single sample function. Fortunately, in practice, random data representing stationary physical phenomena are generally ergodic. It is for this reason that the properties of stationary processes can be measured properly, in most cases, from a single observed time history record.

Tests for Stationarity. Perhaps the simplest way to evaluate the stationarity of sample random data is to consider the physics of the phenomenon that produces the data. If the basic physical factors which generate the phenomenon are time invariant, then the stationarity of the resulting data can generally be accepted without further study. For example, consider the random data representing pressure fluctuations in the turbulent boundary layer generated by the flight of a high speed aircraft. If the aircraft is flying at constant altitude with a constant speed and with a fixed configuration, it would be reasonable to assume that the resulting pressure data are stationary. In practice, however, data are often collected under circumstances which do not permit an assumption of stationarity based on simple physical considerations. In such cases, the stationarity of the data must be evaluated by studies of available sample records. The sample record used in such studies must be long enough to permit non-stationary trends to be differentiated from the random fluctuations of the time history.

If a time series of the process is available, the test to be performed would be whether the properties of the series do or do not change along the series. For example, if the series is divided into two parts and if it is shown that parameters of both the parts are not significantly different among themselves, both parts are considered as being from the same process. If this is repeated for still smaller parts and the same results are obtained, the process is inferred to be stationary. In practice, if the mean and the covariance are inferred as not being dependent on the position along a time series, the series is considered to be second order stationary. In general, if the subseries parameters are confined within 95 tolerance limits about the corresponding value of the parameter for the entire series then the process is inferred to be stationary.

8.4. DESCRIPTION OF A STOCHASTIC PROCESS

The probabilistic structure of a stationary stochastic process can be described in two ways. They are :

(*i*) By specifying the joint probability distribution of n time points $\{X(t_1), X(t_2), ..., X(t_n)\}$ for any arbitrary t_i and n.

(*ii*) By specifying an identical probability distribution function $F(x)$ for all time positions and a dependence function $D(x_t, x_{t-1}, x_{t-2} ......)$ which gives the description of the connections between the successive values.

The first method, in which n-dimensional probability distribution for the sample series of size n is specified, uniquely describes a stochastic process. However, there are practical difficulties in treating such functions and estimating their parameters for two reasons. First, the multivariate distributions in hydrology and similar sciences are rarely Gaussian that can be readily analysed and secondly relations between n variables may be difficult to handle except in the form of matrices because of the large amount of data. These two reasons induced the specialists in applied sciences to attempt a much simpler approach by reducing the process, whenever feasible, to identically distributed variables and suggesting a suitable mathematical model for dependence, before attempting the approach of the n-dimensional multivariate distributions.

The ensuing treatment here will use mainly the latter approach, that is, with an identical distribution and a dependence model. Once these two functions $F(x)$ and $D(x_t, x_{t-1}, x_{t-2}, ...)$ are postulated and their parameters estimated by performing proper statistical tests of hypotheses, the stochastic process will be considered as being mathematically described.

8.5. CLASSIFICATION OF TIME SERIES

A time series which is an observed realisation of the stochastic process is the basis for analysis, identification and description of the process. Any time series of a random phenomenon can generally be considered to consist of a *deterministic component* which is the one that can be determined uniquely and a stochastic component consisting of chance and chance dependent effects. The deterministic series and the stochastic series are subdivided into different groups as given in Fig. 8.3. A brief description of these time series now follows.

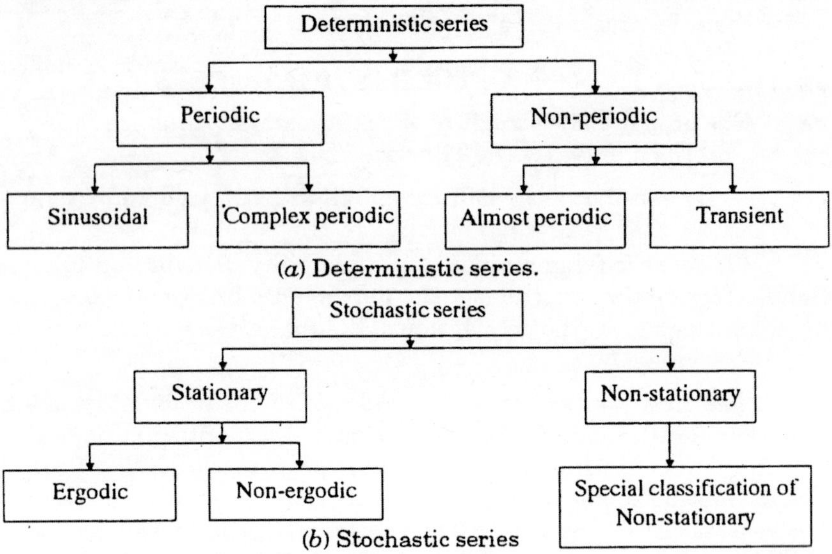

(*a*) Deterministic series.

(*b*) Stochastic series

Fig. 8.3. Classification of time series.

Sinusoidal series. Sinusoidal data are those types of periodic data which can be defined mathematically by a time varying function of the form

$$x(t) = C \sin (2 \pi f_0 t + \phi) \qquad ...(8.7)$$

where C is the amplitude, f_0 is the cyclic frequency in cycles per unit time, ϕ is the initial phase angle with respect to the time origin in radians and $x(t)$ is the instantaneous value at t. The sinusoidal series are shown in Fig. 8.4.

The time interval required for one full fluctuation or for one full cycle is called the *period* and is denoted by T_p. The number of cycles per unit time is called the *frequency* denoted by f_0. The frequency and the period are related by

$$T_p = 1/f_0 \qquad ...(8.8)$$

There are many examples of physical phenomena which produce approximately sinusoidal data in practice. The voltage output of an electric generator is one example. No hydrologic time series is a simple sine or cosine function. However, in some rare cases the deterministic periodic component of monthly rainfall may show approximately a sine function.

Complex periodic series. The basic property of a periodic series is that it repeats itself at regular intervals, T_p, such that

$$x(t) = x(t + nT_p) \; ; \; n = 1, 2, 3, ... \qquad ...(8.9)$$

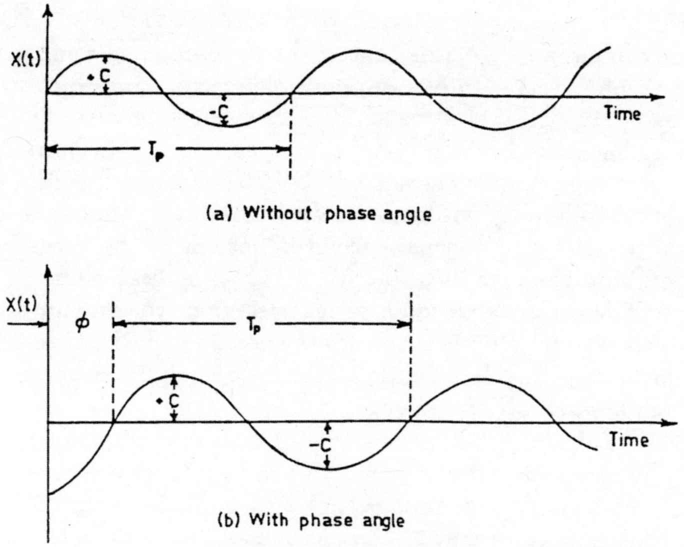

Fig. 8.4. Sinusoidal series.

As in the case of sinusoidal series, the time interval required for one full cycle is called the period. The number of cycles per unit time is called the *fundamental frequency* f_1. A special case of complex periodic series is clearly the sinusoidal series when $f_1 = f_0$. In general, a complex periodic data may be expended into a Fouries series according to the following formula

$$x(t) = \bar{x} + \sum_{n=1}^{\infty} [a_n \cos(2\pi\, n\, f_1\, t) + b_n \sin(2\pi\, n\, f_1\, t)] \qquad ...(8.10)$$

where $f_1 = 1/T_p$ is the fundamental frequency. Alternatively, the complex periodic series can also be expressed in the form

$$x(t) = \bar{x} + \sum_{n=1}^{\infty} C_n \sin(2\pi\, n f_1 t + \phi_n) \qquad ...(8.11)$$

where a_n and b_n are called Fourier coefficients related to C_n and ϕ_n as given below

$$C_n^2 = a_n^2 + b_n^2 \qquad ...(8.12)$$

$$\phi_n = \tan^{-1}(-b_n/a_n) \qquad ...(8.13)$$

In words Eq. (8.11) says that complex periodic data consists of a static component $\bar{x}$ and an infinite number of sinusoidal components called harmonics, which have amplitudes C_n and phases ϕ_n. The frequencies of the harmonic components are all integral multiples of the fundamental frequency f_1.

Sometimes complex periodic data will include only a fewer harmonic components. In other cases, the fundamental component may be absent. For example, suppose a periodic time history is formed by mixing three sine waves which have frequencies of 60, 75 and 100 cycles/second. The highest common divisor is 5. Therefore, the period of the resulting series will be 1/5 = 0.2 seconds and the fundamental frequency will be 5 cycles/second. Note that this fundamental frequency of 5 cycles/second is not one of the three harmonics making the complex periodic series. Physical phenomena which produce complex periodic series are far more common than those which produce simple sinusoidal series. In fact, the classification of series as being sinusoidal is often only an approximation for data which are actually complex.

Nearly all hydrologic time series of daily and monthly values (and of any other interval in to which the year may be divided in order to transform the continuous time series into discrete series of various time intervals) of such phenomena as precipitation, runoff, evaporation, ground water and lake levels, sediment transport, river water qualities etc. have deterministic components which are periodic series. The main cyclic frequency is 1/12 for monthly values (or one cycle in twelve months) and 1/365 (or one cycle in 365 days) for daily values. Similarly, the complex periodic component of one cycle per day is present in many hydrologic series like the discharge affected by melting of snow and ice, evaporation, radiation, temperature, etc., when they are observed with time interval less than a day. However, these complex periodic components are always associated and combined with stochastic components of greater or smaller effects on the series in comparison with periodic components.

Almost periodic series. This type of series is described mathematically by

$$x(t) = \bar{x} + C_1 \cos (2\pi f_1 t + \phi_1) + C_2 \cos (2\pi f_2 t + \phi_2) + \ldots\ldots + C_n \cos (2\pi f_n t + \phi_n) \qquad \ldots(8.14)$$

where (f_i/f_j) is not a rational number in all cases. No common factor exists for all frequencies f_j, such as in the case of complex periodic series. No period T_p can be found for almost periodic data. Physical phenomena producing almost periodic series frequently occur in practice when the effects of two or more unrelated periodic phenomena are mixed. A typical example of a series composed of almost periodic and stochastic elements are tidal levels along coasts and estuaries. The three cycles which affect the tides namely the earth's daily rotation, the moon's revolution around the earth and the earth's revolution around the sun do not have commensurate frequencies.

Transient series. Transient series are defined as all non-periodic data other than the almost periodic data discussed in the above section. In other words, transient data include all data, not previously discussed, which can be described by some suitable time varying function. Three simple examples of transient series are given in Fig. 8.5. Physical phenomena which produce transient series are numerous and diverse. These types of series are only casual components in the case of hydrologic series, superimposed on other components such as periodic and stochastic components. Two simple types of transient components are most commonly detected in hydrology. They are *linear or non-linear trends* and *jumps* as shown in Fig. 8.6. Jumps are created by sudden changes that are man-made or natural.

$$x(t) = Ae^{-at} \quad t \geq 0$$
$$= 0 \quad t < 0$$

$$x(t) = Ae^{-at} \cos bt \quad t \geq 0$$
$$= 0 \quad t < 0$$

$$x(t) = A \quad C \geq t \geq 0$$
$$= 0 \quad \text{elsewhere}$$

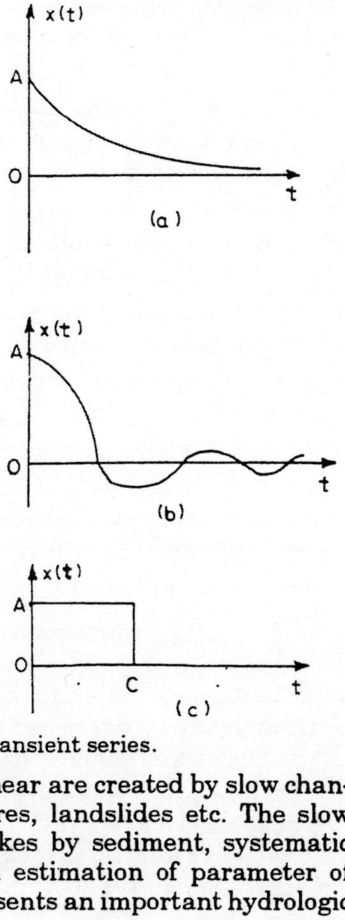

Fig. 8.5. Examples of transient series.

Trends, either linear or non-linear are created by slow changes. The sudden changes include fires, landslides etc. The slow changes may be due to filling of lakes by sediment, systematic afforestation etc. The detection and estimation of parameter of jumps and trends in time series represents an important hydrologic subject.

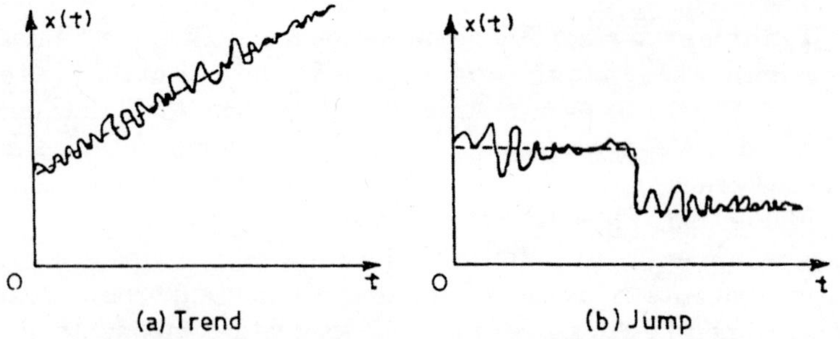

Fig. 8.6. Transient components in hydrologic series.

Stationary series. As already defined, the stationary stochastic series are time invariant. They are important in hydrology mainly for two reasons. First, the mathematical techniques for the analysis of these series are well developed. Secondly in most of the hydrologic series, once the known deterministic components have been identified and removed from the series, the remaining stochastic component may be considered approximately stationary. Most of the hydrologic series are also ergodic enabling the estimation of parameters from a single sample function. The stationary ergodic series can be divided into *dependent* and *independent series*. An independent stationary stochastic series is called a *pure random series* which is completely described by a single probability distribution function. Dependent stationary series in turn can further be divided into *linear dependent series* and *non-linear dependent series*. However, it is only linear dependent series for which thorough techniques of analysis have been developed and which find applications in many physical phenomena. The two basic linear dependent ergodic stationary stochastic series are the *moving average series* and the *autoregressive series*. The linear dependent and the independent series will be discussed in detail in Chapters 9 and 10.

Non-stationary series. These series are also called *evolutive series* or *time variant series* because their various statistical properties change with time. Strictly speaking on a long-range basis nearly all hydrologic series are non-stationary in character because the river basin and climate slowly change with time. However, these changes in the time spans of our interest are relatively small so that hydrologic series are normally considered to be stationary. *Autoregressive integrated moving average series* represent a kind of non-stationary series for which the analysis techniques have been developed.

8.6. COMPONENTS OF TIME SERIES

An observed time series of a random phenomenon may consist of an arbitrary component such as a *catastrophic event*, a deterministic component which may be a non-periodic transient or periodic or a combination of both, and lastly a stochastic component. That is,

$$x(t) = x'(t) + x''(t) + x'''(t) + x''''(t) \quad ...(8.15)$$

where $x(t)$ is the observed time series

$x'(t)$ is the stochastic component

$x''(t)$ is the linear trend

$x'''(t)$ is the catastrophic event

$x''''(t)$ is the periodic or cyclic component.

Again this stochastic component $x'(t)$ may be dependent or independent. If the stochastic component is independent, then it is called a *pure random component*. The dependence that exists in the stochastic component is called the *persistence*. All these components are shown in Fig. 8.7.

The catastrophic or arbitrary events are rare events about which no prediction can be made because of lack of data and their arbitrary behaviour. The trend component may be introduced into the series because of the slow changes that are natural or manmade. The cyclicity in the series may be due to the systematic repetitive nature of factors governing the phenomenon. Hydrologic series, more often than not, are combinations of deterministic and stochastic components. The deterministic component is made of cyclicity corresponding to the annual cycle when the original series are observed with time intervals of less than a year. Now we shall deal with the methods of investigation to identify and quantitatively evaluate these components that are present in the series.

Methods of investigation. Any method which is used to investigate a given hydrological time series should result in the same basic conclusions about the structure of the series. The method should then extract all or most of the information concealed in the observed time series. There are mainly four methods which are practiced. They are :

(*i*) Method of autocorrelation coefficients

(*ii*) Method of spectral analysis

(*iii*) Method of mean ranges

(*iv*) Method of mean runs.

The first two of the above methods are very classical and are the ones which are most commonly employed. They are the subjects of detailed discussion in the next two chapters.

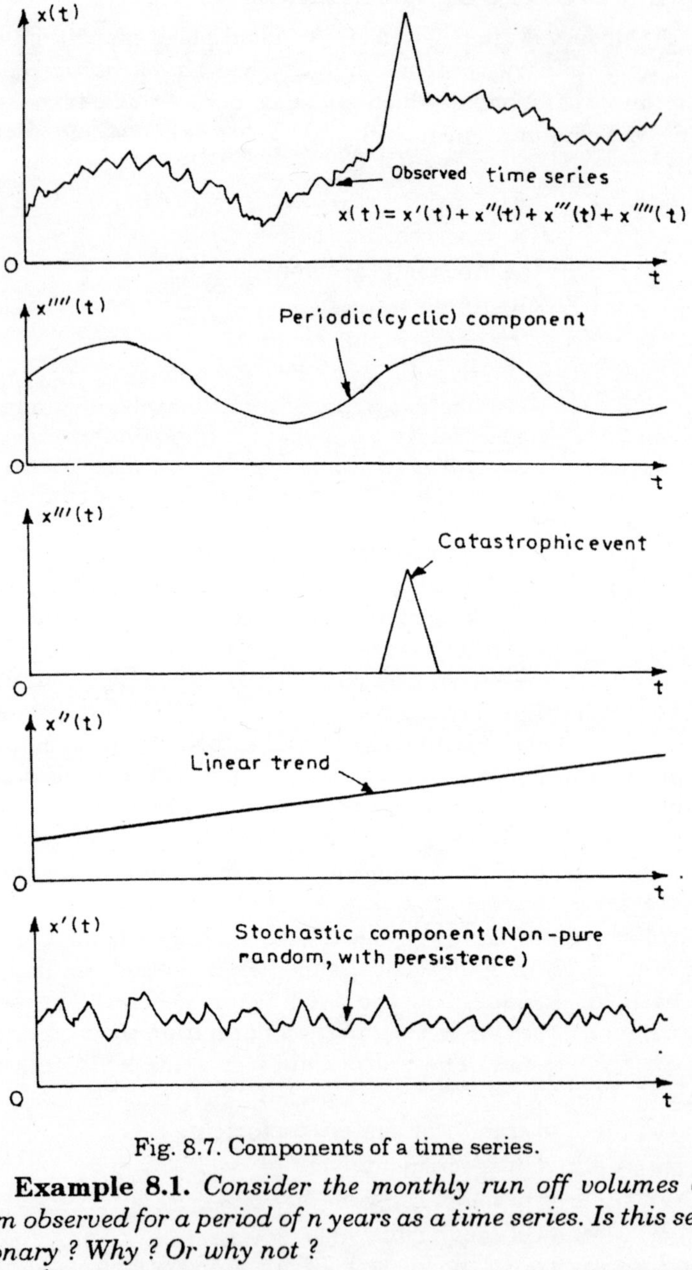

Fig. 8.7. Components of a time series.

Example 8.1. *Consider the monthly run off volumes of a stream observed for a period of n years as a time series. Is this series stationary ? Why ? Or why not ?*

Sol. The run off record of *n* years may be arranged to form an ensemble of *n* series with each series containing monthly run off of

12 month in a year. If the means, variances, and covariances of lag one month and other parameters related to other higher moments are computed for each month, it can be easily noticed that these parameters vary from month to month indicating that the series is not stationary. The obvious reason for this is the presence of the periodic component with a period of one year present in these parameters.

EXERCISES

8.1. Define a stochastic process. Differentiate between stationary and non-stationary processes and ergodic and non-ergodic processes.

8.2. When do you say a process is weakly stationary ? Is a strong stationary process also weakly stationary always ?

8.3. How are stochastic processes classified based on the nature of state and time ?

8.4. What is the difference between a stochastic process and a time series ? What are the various components that can be present in a time series ?

8.5. Distinguish between sinusoidal, complex periodic and almost periodic series. A series is generated by superimposing three periodic components with frequencies of 3, 7 and 11 cycles/second. Is this series complex periodic or almost periodic ?

8.6. Give some examples of hydrologic time series which are stationary and which are non-stationary.

8.7. Test whether the annual run off serires of Example 6.4 is stationary by dividing the total 78 year record into two segments of 39 years each and computing the mean, standard deviation and covariance of lag one year for each segment.

8.8. Repeat Exercise 8.7 by dividing the record into four segments of length 20, 19, 20 and 19 years.

9

Autocorrelation Analysis

9.1. ESTIMATION OF THE AUTOCORRELATION COEFFICIENT

The autocorrelation analysis of random data is a well developed technique. It is very extensively used to determine the linear dependence existing in a series. The method consists of estimating the autocorrelation coefficients for various lags and plotting them against the respective lags resulting in what is known as an empirical correlogram. The empirical correlogram thus obtained is then compared with the theoretical correlograms of known process to identify a suitable mathematical model to describe the process under consideration.

The covariance between $X(t)$ and $X(t + \tau)$ which are separated by a time interval of τ, is called the *autocovariance* at lag τ and is denoted by $\gamma(\tau)$. It is defined by

$$\gamma(\tau) = \text{Cov}\,[X(t), X(t + \tau)] = E[\{X(t) - \mu\}\,\{X(t + \tau) - \mu\}] \quad \text{...(9.1)}$$

where μ is the mean or the expected value of the process. When $\tau = 0$ in Eq. (9.1), we get

$$\begin{aligned}\gamma(0) &= \text{Cov}\,[X(t), X(t)] = E\,[\{X(t) - \mu\}\,\{X(t) - \mu\}]\\ &= E[\{X(t) - \mu\}^2] = \sigma_x^{\,2}\end{aligned}$$

Thus $\gamma(0)$ is nothing but the variance of the process. The autocorrelation coefficient at lag τ is denoted by $\rho(\tau)$ and is given by

$$\rho(\tau) = \gamma(\tau)/\gamma(0) \quad \text{...(9.2)}$$

The autocorrelation coefficient is also referred to as *serial correlation coefficient* and may be abbreviated as either a.c.c. or s.c.c.

If a continuous time stationary stochastic process is assumed to be ergodic, then its autocorrelation coefficient can be obtained as

$$\rho(\tau) = \underset{T \to \infty}{\text{Lt}} \frac{\dfrac{1}{(T - \tau)}\displaystyle\int_0^{T-\tau} \{X(t) - \mu\}\{X(t + \tau) - \mu\}\, dt}{\dfrac{1}{T}\displaystyle\int_0^{T} \{X(t) - \mu\}^2 .\, dt} \quad \text{...(9.3)}$$

where μ is the expected value of the process given by

$$\mu = \underset{T \to \infty}{\text{Lt}} \frac{1}{T} \int_0^T X(t)\, dt \qquad ...(9.4)$$

Similarly for a discrete time stochastic process, the autocorrelation coefficient at lag k is given by

$$\rho_k = \underset{N \to \infty}{\text{Lt}} \frac{\dfrac{1}{(N-k)} \displaystyle\sum_{t=1}^{N-k} (X_t - \mu)(X_{t+k} - \mu)}{\dfrac{1}{N} \displaystyle\sum_{t=1}^{N} (X_t - \mu)^2} \qquad ...(9.5)$$

where μ is the mean of the process given by

$$\mu = \underset{N \to \infty}{\text{Lt}} \frac{1}{N} \sum_{t=1}^{N} X_t \qquad ...(9.6)$$

The autocorrelation coefficients $\rho(\tau)$ or ρ_k given by Eq. (9.3) and Eq. (9.5) and the mean given by Eq. (9.4) and Eq. (9.6) are the population values. In practice, however, we will not be able to know the exact values of these population parameters. We have to estimate them from an observed time series. The values estimated from an observed time series are called the sample mean and sample autocorrelation coefficient and are denoted by $\bar{x}$ and r_k respectively. As we have been mentioning earlier, most of the hydrologic processes are approximated as discrete time continuous state stochastic processes and hence the method of estimating sample autocorrelation coefficients is discussed here only for discrete time series.

From an observed discrete time series x_t, $t = 1, 2, ..., n$, the serial correlation coefficient r_k for any lag k can be obtained by correlating the sequences $(x_1, x_2, ..., x_{n-k})$ and $(x_{1+k}, x_{2+k}, ..., x_n)$ as follows.

$$r_k = \frac{\Sigma (x_t - \bar{x}_t)(x_t - \bar{x}_{t+k})}{\left\{[\Sigma (x_t - \bar{x}_t)^2][\Sigma (x_t - \bar{x}_{t+k})^2]\right\}^{1/2}} \qquad ...(9.7)$$

where all summations run from $t = 1$ to $t = (n - k)$. In Eq. (9.7) $\bar{x}_t$ and $\bar{x}_{t+k}$ are the sample means for the first $(n-k)$ terms and the last $(n-k)$ terms and are given by $\bar{x}_t = \displaystyle\sum_{t=1}^{n-k} \frac{x_t}{(n-k)}$ and $\bar{x}_{t+k} = \displaystyle\sum_{t=1}^{n-k} \frac{x_{t+k}}{(n-k)}$. This is based on the well known definition of correlation coefficient. For computational convenience Eq. (9.7) can also be written in the form

$$r_k = \frac{\Sigma x_t x_{t+k} - (\Sigma x_t \Sigma x_{t+k})/(n-k)}{\left[\Sigma x_t^2 - (\Sigma x_t)^2/(n-k)\right]^{1/2} \left[\Sigma x^2_{t+k} - (\Sigma x_{t+k})^2/(n-k)\right]^{1/2}} \quad ...(9.8)$$

where all the summations again run from $t = 1$ to $t = (n - k)$. Both Eq. (9.7) and Eq. (9.8) are based on the scatter diagram of x_{t+k} and x_t for $t = 1, 2, ..., (n-k)$. Thus as the value of k increases, for a given n, the pairs of observations available for estimating r_k decrease and thereby making the estimate increasingly unreliable. Therefore, it is often suggested that the maximum lag upto which the autocorrelation coefficients can be estimated may be limited to $n/4$ or $n/3$.

If the autocorrelation coefficients ρ_k are plotted against their respective lags k, the resulting graph is called the *autocorrelation function* or the *theoretical correlogram*. Whereas the plot of sample autocorrelation coefficient r_k against k is called the *empirical correlogram*. Sometimes instead of ρ_k the autocovariance $\gamma(k)$ is plotted against the lag k. In such a case the resulting graph is called the *autocovariance function*. It may be observed here that the autocorrelation function of continuous time process is also continuous and likewise the autocorrelation function of discrete time process is also discrete. Some properties of the autocorrelation function are summarised below :

(*i*) $\rho_0 = 1$. This can be easily verified from $\rho_k = \gamma(k)/\gamma(0)$ by letting $k = 0$.

(*ii*) $-1 \leq \rho_k \leq 1$. This result also is very obvious since the autocorrelation coefficient is just like any other correlation coefficient and therefore has to lie between -1 and 1.

(*iii*) For a stationary process $\rho_k = \rho_{-k}$. This follows from the definition of the stationary process. We have Cov (k) = Cov $[X_t, X_{t+k}]$.

In a stationary process, since covariance depends only on the lag time separating the variables and not on the absolute time we can write Cov (k) = Cov $[X_{t-k}, X_t]$. Or Cov (k) = Cov $[X_t, X_{t-k}]$ = Cov $(-k)$. Therefore for a stationary process it is sufficient if we estimate the autocorrelation coefficients for positive lags.

9.2. CORRELOGRAM OF AN INDEPENDENT PROCESS

Consider an independent process Z_t, $t = 1, 2, ...$ Then for this process Cov (k) = Cov $[Z_t, Z_{t-k}] = 0$ since Z_t and Z_{t-k} are independent of each other for any k except for $k = 0$. Since Cov $(k) = 0$, ρ_k must also be zero for all k except for $k = 0$. Obviously $\rho_k = 1$ when $k = 0$.

Therefore, the theoretical correlogram of an independent process must have an ordinate equal to 1 at $k = 0$ and an ordinate of zero for $k > 0$.

But when estimated from an observed independent time series z_t, $t = 1, 2, \ldots\ldots, n$, the autocorrelation coefficient r_k will not be exactly equal to zero. Still we can accept the series to be an independent series if it can be established that the values of r_k, though physically different from zero are statistically indistinguishable from zero. To verify this fact the sampling distribution of r_k is required. Anderson has proposed a test of significance for the autocorrelation coefficient of a *circular*, normal, stationary time series. A circular series is one that closes on itself so that x_n is followed by x_1. Although the assumption of circular series is unrealistic, values of r_k obtained for circular series will not differ greatly from those calculate for the original series if n is large in comparison to k. Under these conditions r_k will be approximately normally distributed with mean equal to $-1/(n-1)$ and variance equal to $(n-2)/(n-1)^2$. The tolerance limits of r_1, for 5% level of significance are then given by

$$r_{1\,\text{upper}} = \frac{-1 + 1.96\sqrt{n-2}}{(n-1)} \quad \ldots(9.9)$$

$$r_{1\,\text{lower}} = \frac{-1 - 1.96\sqrt{n-2}}{(n-1)} \quad \ldots(9.10)$$

The tolerance limits for other correlation coefficients are obtained by replacing n in Eqs. (9.9) and (9.10) by $(n - k + 1)$. Thus

$$r_{k\,\text{upper}} = \frac{-1 + 1.96\sqrt{n-k-1}}{(n-k)} \quad \ldots(9.11)$$

$$r_{k\,\text{lower}} = \frac{-1 - 1.96\sqrt{n-k-1}}{(n-k)} \quad \ldots(9.12)$$

Similarly if the tolerance limits are to be obtained for 1% significance level the constants 1.96 in Eqs. (9.9) through (9.12) is to be replaced by 2.57. It has been shown that the difference between the approximate tolerance limits obtained from the above equations and exact values obtained without assuming the series to be circular is relatively very small. Hence these equations are generally used to test any series for independence.

The procedure to test a given series for independence then consists of plotting the empirical correlogram and on the same diagram constructing the tolerance band with upper and lower tolerance limits. If all the estimated autocorrelation coefficients for various lags are falling within the tolerance band, the hypothesis that the sample function is from an independent process is accepted. Example 9.1 demonstrates this procedure.

When the empirical correlogram is found not to represent an independent process, further analysis is to be carried out to identify from which of the dependent processes the sample function has been drawn. The procedure followed to achieve this objective is to compare the empirical correlogram of the sample function with the theoretical correlograms of some standard processes. The properties of the theoretical correlograms of some of the important processes are discussed in the sections that follow.

Example 9.1. *The 78 year annual run off data of Example 6.4 is used to obtain the first 10 autocorrelation coefficients which are given below. Test the series for independence at 5% level of significance.*

lag k in years	1	2	3	4	5	6	7	8	9	10
r_k	0.27	0.18	0.22	0.00	0.12	0.24	0.22	0.05	0.00	0.00

Sol. The autocorrelation function is plotted in Fig. 9.1. A visual inspection of the correlogram reveals that all the correlation coefficients are either positive or zero and they are not fluctuating around zero. If the series is independent any deviation of r_k from zero is only due to sampling errors and hence we expect r_k to fluctuate around zero. Therefore at this stage we may conclude qualitatively that the process is not independent.

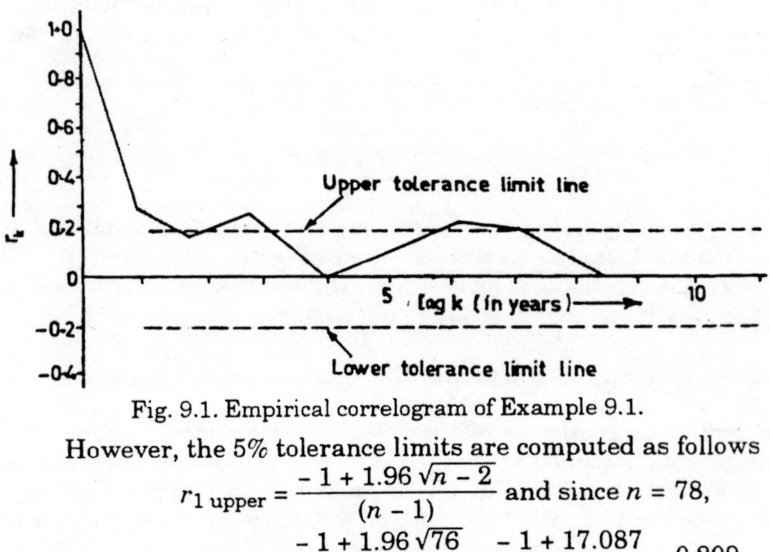

Fig. 9.1. Empirical correlogram of Example 9.1.

However, the 5% tolerance limits are computed as follows

$$r_{1\,\text{upper}} = \frac{-1 + 1.96\sqrt{n-2}}{(n-1)} \text{ and since } n = 78,$$

$$r_{1\,\text{upper}} = \frac{-1 + 1.96\sqrt{76}}{77} = \frac{-1 + 17.087}{77} = 0.209$$

and $$r_{1\,\text{lower}} = \frac{-1 - 17.087}{77} = -0.235.$$

Similarly the upper and lower tolerance limits of r_{10} are computed to be 0.221 and – 0.251 respectively. The tolerance limits for the other autocorrelation coefficients in between are obtained by joining the limits of r_1 and r_{10} by straight lines. The upper tolerance limit line and the lower tolerance limit line together form the tolerance band. The tolerance band is constructed on the same diagram as shown is Fig. 9.1. Since many of the autocorrelation coefficients are falling outside the tolerance band the annual run off series being analysed is not independent.

9.3. MOVING AVERAGE PROCESS

The mth order moving average process abbreviated as $MA(m)$ process is given by

$$X_t = \xi_t - \beta_1 \xi_{t-1} - \beta_2 \xi_{t-2} - \ldots\ldots - \beta_m \xi_{t-m} \quad \ldots(9.13)$$

where β_j, $j = 1, 2, \ldots\ldots, m$ are called the parameters of the process and ξ_t is a pure random process. The process X_t is obtained as a weighted sum of a finite number of previous values of an independent process. The name "moving average" is somewhat misleading because the weights $\beta_1, \beta_2, \ldots\ldots, \beta_n$ need not add up to 1 nor need they be positive always. Taking expectations of Eq. (9.13) on both sides

$$E[X_t] = E[\xi_t] - \beta_1 E[\xi_{t-1}] - \ldots\ldots - \beta_m E[\xi_{t-m}]$$

$$\mu_x = \mu_\xi - \beta_1 \mu_\xi - \ldots\ldots - \beta_m \mu_\xi$$

$$\mu_x = \mu_\xi \left[1 - \sum_{j=1}^{m} \beta_j \right] \quad \ldots(9.14)$$

If the process X_t is to have the same expected value as ξ_t, that is if μ_x is to be equal to μ_ξ, the required condition is that $\sum_{j=1}^{m} \beta_j = 0$.

Alternatively we can define the $MA(m)$ process as

$$(X_t - \mu) = \xi_t - \beta_1 \xi_{t-1} - \beta_2 \xi_{t-2} - \ldots\ldots - \beta_m \xi_{t-m} \quad \ldots(9.15)$$

In this case the expected value of ξ_t must be zero, that is, $\mu_\xi = 0$ and the sum of the parameter β_j need not be equal to zero. Henceforth we shall represent the $MA(m)$ process by Eq. (9.15) only. Squaring Eq. (9.15) on both sides

$$(X_t - \mu)^2 = (\xi_t - \beta_1 \xi_{t-1} - \ldots\ldots - \beta_m \xi_{t-m})(\xi_t - \beta_1 \xi_{t-1} - \ldots\ldots - \beta_m \xi_{t-m})$$

$$(X_t - \mu)^2 = \xi_t^2 + \beta_1^2 \xi_{t-1}^2 + \beta_2 \xi_{t-2}^2 + \ldots\ldots + \beta_m^2 \xi_{t-m}^2 - \{\text{terms like } \beta_i \beta_j \xi_{t-i} \xi_{t-j} \text{ where } i \neq j\} \quad \ldots(9.16)$$

The expectation of $\{\xi_{t-i}\, \xi_{t-j}$, when $i \neq j\}$ will be zero since ξ_{t-i} and ξ_{t-j} are independent when $i \neq j$. Similarly the expectation of ξ_{t-1}^2 will be equal to the variance of the independent process ξ_t as ξ_t has an expected value of zero. Let the variance of ξ_t be denoted by $\sigma_\xi^{\ 2}$. Taking the expectations on both sides of Eq. (9.16) will therefore give

$$\sigma_x^{\ 2} = \sigma_\xi^{\ 2} + \beta_1^{\ 2}\,\sigma_\xi^{\ 2} + \ldots\ldots + \beta_m^2\,\sigma_\xi^{\ 2}$$

or
$$\sigma_x^{\ 2} = (1 + \beta_1^{\ 2} + \beta_2^{\ 2} + \ldots\ldots + \beta_m^2)\,\sigma_\xi^{\ 2} \qquad \ldots(9.17)$$

where $\sigma_x^{\ 2}$ is the variance of the process X_t. Now multiplying Eq. (9.15) on both sides by $(X_{t-k} - \mu)$, we get

$$(X_t - \mu)\,(X_{t-k} - \mu) = (\xi_t - \beta_1\,\xi_{t-1} - \ldots\ldots - \beta_m\,\xi_{t-m})\,(\xi_{t-k} - \beta_1\,\xi_{t-1-k} - \ldots\ldots - \beta_m\,\xi_{t-m-k})$$

Taking the expectations on both sides of the above equation, we get

$$\text{Cov}\,(k) = \sigma_\xi^{\ 2}\left(-\beta_k + \sum_{j=1}^{m} \beta_j\,\beta_{j+k}\right)$$

But whenever the suffix of the parameter β exceeds m, its value is zero. Therefore in the above expression when $j > (m - k)$ the terms in the summation will become zero and hence we can restrict the upper limit of the summation to $(m - k)$.

$$\text{Cov}\,(k) = \sigma_\xi^{\ 2}\left(-\beta_k + \sum_{j=1}^{m-k} \beta_j\,\beta_{j+k}\right)$$

$$\text{Cov}\,(k) = \sigma_\xi^{\ 2}\,(-\beta_k + \beta_1\,\beta_{k+1} + \beta_2\,\beta_{k+2} + \ldots\ldots + \beta_{m-k}\,\beta_m) \qquad \ldots(9.18)$$

Using the results of Eq. (9.17) and Eq. (9.18), the autocorrelation coefficient is now given by

$$\rho_k = \frac{\text{Cov}\,(k)}{\text{Cov}\,(0)} = \frac{\sigma_\xi^{\ 2}\,(-\beta_k + \beta_1\,\beta_{k+1} + \beta_2\,\beta_{k+2} + \ldots\ldots\ \beta_{m-k}\,\beta_m)}{\sigma_\xi^{\ 2}\,(1 + \beta_1^{\ 2} + \beta_2^{\ 2} + \ldots\ldots + \beta_m^2)}$$

$$\rho_k = \frac{(-\beta_k + \beta_1\,\beta_{k+1} + \beta_2\,\beta_{k+2} + \ldots\ldots + \beta_{m-k}\,\beta_m)}{(1 + \beta_1^{\ 2} + \beta_2^{\ 2} + \ldots\ldots + \beta_m^2)} \qquad \ldots(9.19)$$

Evidently $\rho_k = 0$ for $k > m$, since in this case all βs in the numerator of Eq. (9.19) are zero. Thus we see that the autocorrelation function of *MA*(m) process is zero beyond m. In other words, the autocorrelation function of a *MA*(m) process has a *cut-off* at lag m.

Eq. (9.19) gives the ordinates of the theoretical correlogram of a *MA*(m) process. As an example consider a *MA*(2) process with $\beta_1 = \beta_2 = -\frac{1}{2}$. Then we have $1 + \beta_1^{\ 2} + \beta_2^{\ 2} = 1 + 0.25 + 0.25 = 1.5$. And

$$\rho_1 = \frac{(-\beta_1 + \beta_1\beta_2)}{1.5} = \frac{[0.5 + (-0.5) \times (-0.5)]}{1.5} = 0.5$$

$$\rho_2 = -\frac{\beta_2}{1.5} = -\frac{(-0.5)}{1.5} = \frac{1}{3} = 0.333$$

and $\rho_k = 0$ for $k \geq 3$.

The correlogram of this process is shown in Fig. 9.2.

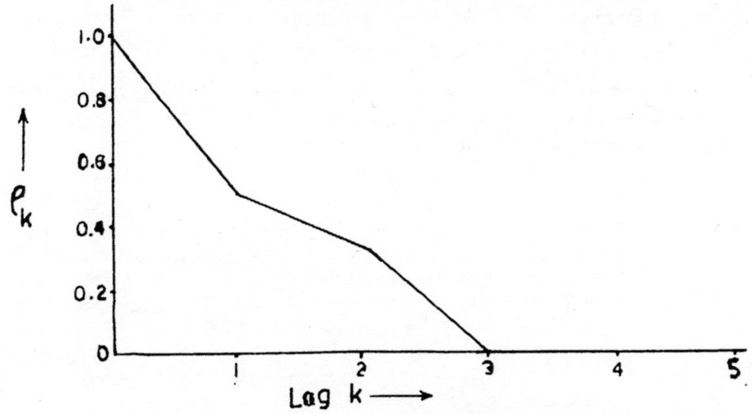

Fig. 9.2. Theoretical correlogram of second order M.A. process.

Moving average models as such have found little application in describing the hydrologic data. However, when they are jointly considered with autoregressive models a new kind of models such as *autoregressive moving average* (ARMA) models or *autoregressive integrated moving average* (ARIMA) models are developed which have wider application.

Example 9.2. *A moving average process is given by* $(X_t - \mu) = \xi_t - 1.3\,\xi_{t-1} + 0.4\,\xi_{t-2}$. *Determine the first four autocorrelation coefficients of the process. Also find the variance of X given variance of* ξ *is 1.0.*

Sol. Comparing the given process with Eq. (9.15), we notice that it is a $MA(2)$ process with $\beta_1 = 1.3$, $\beta_2 = -0.4$ and $m = 2$. Therefore $1 + \beta_1^2 + \beta_2^2 = 1 + (1.3)^2 + (-0.4)^2 = 2.85$. The correlation coefficients are given by

$$\rho_1 = \frac{-\beta_1 + \beta_1\,\beta_2}{2.85} = \frac{-1.3 + (1.3 \times -0.4)}{2.85} = \frac{-1.82}{2.85} = -0.64$$

$$\rho_2 = \frac{-\beta_2}{2.85} = \frac{-(-0.4)}{2.85} = 0.14$$

and $\rho_k = 0$ for $k \geq 3$.

Therefore the first 4 autocorrelation coefficients are – 0.64, 0.14, 0.0 and 0.0.

From Eq. (9.17), we have

$$\sigma_x^2 = (1 + \beta_1^2 + \beta_2^2)\, \sigma_\xi^2$$

$$\sigma_x^2 = 2.85 \times 1 = 2.85.$$

Example 9.3. *The first serial correlation coefficient of MA(1) process is 0.4. Determine the parameter of the process.*

Sol. From Eq. (9.19), the relation between the parameter of the process β_1 and the first serial correlation coefficient ρ_1 for $MA(1)$ process is given by

$$\rho_1 = \frac{-\beta_1}{1 + \beta_1^2}$$

or

$$\rho_1 \beta_1^2 + \beta_1 + \rho_1 = 0$$

$$\beta_1 = \frac{-1 \pm \sqrt{1 - 4\rho_1^2}}{2\rho_1} = \frac{-1 \pm \sqrt{1 - 4 \times (0.4)^2}}{2 \times 0.4}$$

$$= \frac{-1 \pm 0.6}{0.8} = -0.5 \text{ or } -2.0$$

Therefore, the process may be written as

$$(X_t - \mu) = \xi_t + 0.5\, \xi_{t-1}$$

or

$$(X_t - \mu) = \xi_t + 2.0\, \xi_{t-1}.$$

9.4. AUTO REGRESSIVE PROCESS

The mth order autoregressive process, abbreviated as $AR(m)$ process is given by

$$(X_t - \mu_x) = \alpha_1 (X_{t-1} - \mu_x) + \alpha_2 (X_{t-2} - \mu_x) + \ldots\ldots + \alpha_m (X_{t-m} - \mu_x) + \xi_t \quad \ldots(9.20)$$

where μ_x is the expected value of the process, ξ_t is an independent process with mean zero, and $\alpha_1, \alpha_2, \ldots\ldots, \alpha_m$ are the parameters of the process. Multiplying Eq. (9.20) by $(X_{t-k} - \mu_x)$ on both sides and taking expectations

$$E[(X_t - \mu_x)(X_{t-k} - \mu_x)] = \alpha_1 E[(X_{t-1} - \mu_x)(X_{t-k} - \mu_x)] + \alpha_2 E[(X_{t-2} - \mu_x)(X_{t-k} - \mu_x)] + \ldots\ldots + \alpha_m E[(X_{t-m} - \mu_x)(X_{t-k} - \mu_x)] + E[\xi_t (X_{t-k} - \mu_x)] \quad \ldots(9.21)$$

In Eq. (9.21) $E[\xi_t (X_{t-k} - \mu_x)] = 0$, since ξ_t and X_{t-k} are independent whenever $k > 0$. Also $E[(X_{t-j} - \mu_x)(X_{t-k} - \mu_x)]$ is nothing but Cov $(X_{t-j}, X_{t-k}) = \gamma(k - j)$. Therefore Eq. (9.21) may be written as, for $k > 0$,

$$\gamma(k) = \alpha_1 \gamma(k-1) + \alpha_2 \gamma(k-2) + \ldots\ldots + \alpha_m \gamma(k-m) \quad \ldots(9.22)$$

Dividing Eq. (9.22) through out by $\gamma(0)$, we get

$$\rho_k = \alpha_1 \rho_{k-1} + \alpha_2 \rho_{k-2} + \ldots\ldots + \alpha_m \rho_{k-m} \quad \ldots(9.23)$$

In Eq. (9.21), if we make $k = 0$, we get

$$\gamma(0) = \alpha_1 \gamma(1) + \alpha_2 \gamma(2) + \ldots\ldots + \alpha_m \gamma(m) + E[\xi_t (X_t - \mu_x)] \quad \ldots(9.24)$$

But
$$E[\xi_t (X_t - \mu_x)] = E[\xi_t \{\alpha_1 (X_{t-1} - \mu_x) + \alpha_2 (X_{t-2} - \mu_x) + \ldots\ldots + \alpha_m (X_{t-m} - \mu_x) + \xi_t\}]$$
$$= \alpha_1 E[\xi_t (X_{t-1} - \mu_x)] + \alpha_2 E[\xi_t (X_{t-2} - \mu_x)] + \ldots\ldots + \alpha_m E[\xi_t (X_{t-m} - \mu_x)] + E[\xi_t \cdot \xi_t]$$

Since ξ_t is having zero mean $E[\xi_t \cdot \xi_t]$ is equal to the variance of ξ_t, that is σ_ξ^2. Also $E[\xi_t (X_{t-j} - \mu_x)] = 0$ wherever $j > 1$ as ξ_t and X_{t-j} will be then independent of each other. Therefore, we have

$$E[\xi_t (X_t - \mu_x)] = 0 + 0 + 0 + \ldots\ldots + 0 + \sigma_\xi^2$$

And Eq. (9.24) can now be written as

$$\gamma(0) = \alpha_1 \gamma(1) + \alpha_2 \gamma(2) + \ldots\ldots + \alpha_m \gamma(m) + \sigma_\xi^2$$

Dividing the above equation by $\gamma(0)$ throughout, we get

$$1 = \alpha_1 \rho_1 + \alpha_2 \rho_2 + \ldots\ldots + \alpha_m \rho_m + \frac{\sigma_\xi^2}{\gamma(0)}$$

Since $\gamma(0) = \sigma_x^2$, the variance of X_t, we obtain

$$\frac{\sigma_\xi^2}{\sigma_x^2} = (1 - \alpha_1 \rho_1 - \alpha_2 \rho_2 - \ldots\ldots - \alpha_m \rho_m)$$

or

$$\sigma_x^2 = \frac{\sigma_\xi^2}{(1 - \alpha_1 \rho_1 - \alpha_2 \rho_2 - \ldots\ldots - \alpha_m \rho_m)} \quad \ldots(9.25)$$

Eq. (9.25) relates the variance of the independent process σ_ξ^2 to the variance of the *AR* (m) process σ_x^2.

If we substitute $k = 1, 2, \ldots\ldots, m$ in Eq. (9.23), realising that $\rho_{-j} = \rho_j$ for a stationary process, we obtain a set of linear equations for $\alpha_1, \alpha_2, \ldots\ldots, \alpha_m$ in terms of $\rho_1, \rho_2, \ldots\ldots, \rho_m$. That is

$$\left.\begin{aligned}
\rho_1 &= \alpha_1 + \alpha_2\rho_1 + \ldots\ldots + \alpha_m \rho_{m-1} \\
\rho_2 &= \alpha_1\rho_1 + \alpha_2 + \ldots\ldots \alpha_m \rho_{m-2} \\
&\vdots \qquad \vdots \qquad \vdots \\
\rho_m &= \alpha\rho_{m-1} + \alpha_2 \rho_{m-2} + \ldots\ldots + \alpha_m
\end{aligned}\right\} \quad \ldots.(9.26)$$

The equations in (9.26) are called the *Yule-Walker* equations. We can obtain the estimates of the parameters $\alpha_1, \alpha_2, \ldots\ldots, \alpha_m$ by replacing the population autocorrelation coefficients ρ_k with the sample correlation coefficients r_k and solving the m simultaneous equations of (9.26).

First order AR process. The first order *AR* process is given by

$$(X_t - \mu) = \alpha_1(X_{t-1} - \mu_x) + \xi_t \qquad ...(9.27)$$

Letting $m = 1$ in Eq. (9.23), the autocorrelation function of *AR*(1) process is given by

$$\rho_k = \alpha_1 \rho_{k-1}, \text{ for } k > 0 \qquad ...(9.28)$$

When $k = 1$ in this equation we have $\rho_1 = \alpha_1 \rho_0$. Since $\rho_0 = 1$, it means that $\alpha_1 = \rho_1$. In other words, the parameter α_1 of the *AR* (1) process is equal to the first serial correlation coefficient of the process, ρ_1, itself. Now Eq. (9.28) reduces to

$$\rho_k = \rho_1^{\,k} \qquad ...(9.29)$$

The correlogram of the *AR*(1) process is completely described once ρ_1 is known. Since, for a stationary process $\rho_{-k} = \rho_k$, the autocorrelation function is even, that is, symmetrical about the vertical axis. The most general form of Eq. (9.29) would be

$$\rho_k = \rho_1^{|k|} \text{ for } k = 0, \pm 1, \pm 2, \ldots\ldots \qquad ...(9.30)$$

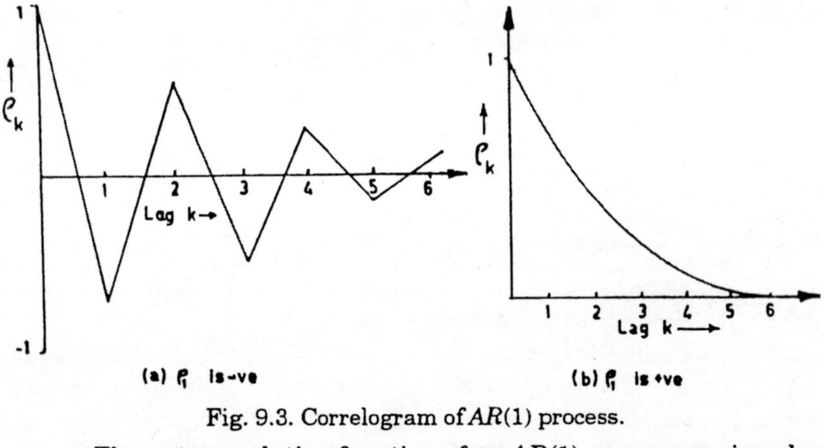

Fig. 9.3. Correlogram of *AR*(1) process.

The autocorrelation function of an *AR*(1) process as given by Eq. (9.30) for the cases of ρ_1 being negative and ρ_1 being positive is shown in Fig. 9.3. As seen from this figure the function decays exponentially to zero when ρ_1 is positive. When ρ_1 is negative the autocorrelation function is a *zig-zag* broken line oscillating about the horizontal axis, again decaying exponentially to zero.

Usually the hydrologic series following the *AR*(1) process will have positive ρ_1. If it is assumed that the flow recession curve of the annual run off at the end of every water year may be well approximated by an exponential curve, it can be shown that the annual flow series is a dependent series that can be described by *AR*(1) model. In fact it has been observed that the annual run off series of many large rivers in the world may be well fitted by $AR(1)$ models with positive ρ_1.

If ρ_1 is to be negative, the series itself has to be *zig-zag* since the adjacent values in such a series are negatively correlated. For any hydrologic variable to be fitted by $AR(1)$ model with negative ρ_1, there must be import and export of that variable in successive time periods which can be ruled out in nature. Therefore $AR(1)$ models with negative ρ_1 do not find any applications in hydrology.

Example 9.4. *The first ten serial correlation coefficients computed from 40 years of annual run off series of a river are given below. Test the series for independence at 5% level of significance and suggest a suitable model to describe the series.*

lag k in years	*1*	*2*	*3*	*4*	*5*	*6*	*7*	*8*	*9*	*10*
r_k	*0.62*	*0.40*	*0.18*	*0.14*	*0.03*	*– 0.02*	*0.07*	*0.01*	*0.00*	*– 0.02*

Sol. The given autocorrelation coefficients are plotted to provide an empirical correlogram as shown in Fig. 9.4. It is obvious from this figure that the annual run off series is not independent.

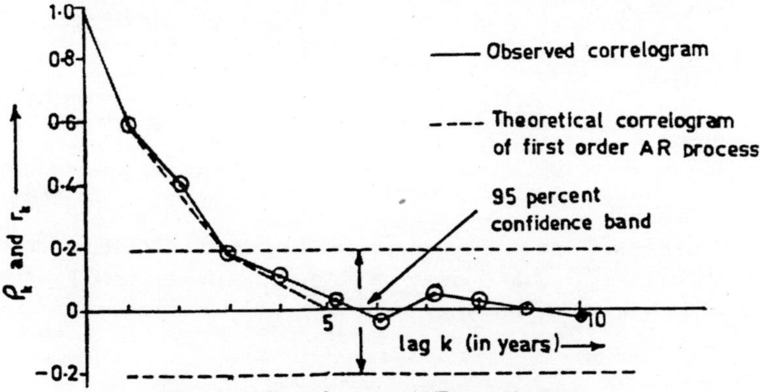

Fig. 9.4. Correlogram of Example 9.4.

However, the tolerance limits are computed for 5% level of significance using Eqs. (9.9) to (9.12) with $n = 40$ for r_1 and r_{10}.

$$\text{upper limit of } r_1 = \frac{-1 + 1.96\sqrt{38}}{39} = 0.284$$

$$\text{lower limit of } r_1 = \frac{-1 - 1.96\sqrt{38}}{39} = -0.335$$

$$\text{upper limit of } r_{10} = \frac{-1 + 1.96\sqrt{29}}{30} = 0.318$$

$$\text{lower limit of } r_{10} = \frac{-1 - 1.96\sqrt{29}}{30} = -0.385.$$

The confidence band is constructed with these values on the same diagram. Since r_1 and r_2 are very much outside the confidence band, the series is not independent. On keen observation it may be noticed that the empirical correlogram is closely resembling the theoretical correlogram of AR (1) process which is computed from $\rho_k = \rho_1^k$ and plotted on the same diagram. Hence an AR(1) model may be taken to be a good choice to fit the given annual run off series.

Second order AR process. The second order AR process is represented by

$$(X_t - \mu_x) = \alpha_1 (X_{t-1} - \mu_x) + \alpha_2 (X_{t-2} - \mu_x) + \xi_t \quad \text{...(9.31)}$$

where ξ_t is an independent process with mean zero. Letting $m = 2$ in Eq. (9.23), the autocorrelation function of AR(2) process is given by

$$\rho_k = \alpha_1 \rho_{k-1} + \alpha_2 \rho_{k-2} \quad \text{...(9.32)}$$

The corresponding Yule-Walker equations for $k = 1$ and $k = 2$ are

$$\left.\begin{aligned} \rho_1 &= \alpha_1 + \alpha_2 \rho_1 \\ \rho_2 &= \alpha_1 \rho_1 + \alpha_2 \end{aligned}\right\} \quad \text{...(9.33)}$$

The solution of these two simultaneous equations for α_1 and α_2 yield

$$\alpha_1 = \frac{\rho_1 - \rho_1\rho_2}{1 - \rho_1^2} \quad \text{...(9.34)}$$

$$\alpha_2 = \frac{\rho_2 - \rho_1^2}{1 - \rho_1^2} \quad \text{...(9.35)}$$

Using r_1 and r_2 in place of ρ_1 and ρ_2 in the above equations, the estimates of α_1 and α_2 may be obtained. For the AR(2) process to be stationary the following conditions must be satisfied by α_1 and α_2.

$$\left.\begin{aligned} \alpha_2 + \alpha_1 &< 1 \\ \alpha_2 - \alpha_1 &< 1 \\ -1 < \alpha_2 &< 1 \end{aligned}\right\} \quad \text{...(9.36)}$$

Likewise, for stationarity the corresponding conditions to be satisfied by ρ_1 and ρ_2 are given by

$$\left.\begin{aligned} -1 < \rho_1 &< 1 \\ -1 < \rho_2 &< 1 \\ \rho_1^2 &< \tfrac{1}{2}(\rho_2 + 1) \end{aligned}\right\} \quad \text{...(9.37)}$$

Having obtained the values of α_1 and α_2 using known values of ρ_1 and ρ_2 in Eqs. (9.34) and (9.35), the successive ordinates of the autocorrelation function of $AR(2)$ process may be obtained by solving Eq. (9.32) successively for $k = 3, 4, \ldots\ldots$

The shape of the autocorrelation function depends on the nature of roots of the characteristic equation $1 - \alpha_1 x - \alpha_2 x^2 = 0$. When the roots of this equation x_1 and x_2 are real, that is when $\alpha_1^2 + 4\alpha_2 \geq 0$, the autocorrelation function consists of mixture of damped exponentials. Specifically, when both the roots are positive the autocorrelation function remains positive as it damps out. On the other hand when one of the roots is negative the autocorrelation function alternates in sign as it damps out. If the roots of the equations are complex, that is when $\alpha_1^2 + 4\alpha_2 < 0$, the autocorrelation function will be in the form of a *damped sine wave*. In this case the autocorrelation function can also be expressed as

$$\rho_k = \frac{\{\operatorname{sgn}(\alpha_1)\}^k d^k \sin(2\pi f_0 k + F)}{\sin F} \qquad \ldots(9.38)$$

where sgn (α_1) is + 1 if α_1 is positive and − 1 if α_1 is negative. In either case Eq. (9.38) is referred to as a damped sine wave with damping factor d, *frequency* f_0 and phase F. These factors are related to the process parameters as given below

$$d = \sqrt{-\alpha_2} \qquad \ldots(9.39)$$

where the positive square root is taken,

$$\cos 2\pi f_0 = \frac{|\alpha_1|}{2\sqrt{-\alpha_2}} \qquad \ldots(9.40)$$

$$\tan F = \frac{(1 + d^2)\tan 2\pi f_0}{(1 - d^2)} \qquad \ldots(9.41)$$

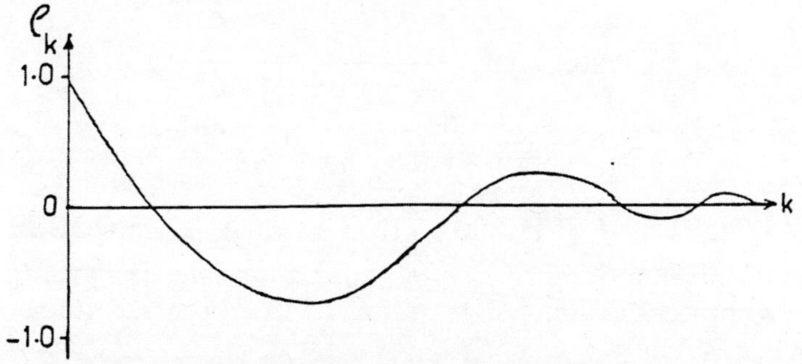

Fig. 9.5. Theoretical correlogram of AR (2) process exhibiting damped sine wave.

The autocorrelation function of an $AR(2)$ process exhibiting a damped sine wave is shown in Fig. 9.5. Sometimes the oscillations in such a correlogram may be misinterpreted as a periodic component.

Therefore one must be careful in comparing the empirical correlogram with the theoretical correlograms of standard processes in such cases.

Expressions for the three parameters of an $AR(3)$ process may be obtained in terms of ρ_1, ρ_2 and ρ_3 from Yule-Walker equations on similar lines as those applied in $AR(2)$ process. But the use of AR (3) model is very rare in hydrology. Mostly $AR(1)$ and occasionally $AR(2)$ models find applications in hydrology.

Example 9.5. *A process is given by* $(X_t - \mu_x) = 1.2\,(X_{t-1} - \mu_x) - 0.5\,(X_{t-2} - \mu_x) + \xi_t$. *Is this process stationary. Determine the variance of* X_t *given the variance of* ξ_t *as 1.0. Also obtain the first 5 serial correlation coefficients.*

Sol. Comparing the given process with Eq. (9.31) we notice that it is an $AR(2)$ process with $\alpha_1 = 1.2$ and $\alpha_2 = 0.5$

$$\alpha_1 + \alpha_2 = 1.2 - 0.5 = 0.7\ ;\ \text{so}\ (\alpha_1 + \alpha_2) < 1$$

$$\alpha_2 - \alpha_1 = -0.5 - 1.2 = -1.7\ ;\ \text{so}\ (\alpha_2 - \alpha_1) < 1$$

$$\alpha_2 = -0.5\ ;\quad \text{so}\ -1 < \alpha_2 < 1$$

Since the parameters α_1 and α_2 satisfy Eq. (9.36) the process is stationary. From Eq. (9.33),

$$\rho_1 = \alpha_1 + \alpha_2\rho_1 = 1.2 + (-0.5)\,\rho_1 \quad \text{or} \quad \rho_1 = 0.8$$

$$\rho_2 = \alpha_1\rho_1 + \alpha_2 = 1.2 \times 0.8 + (-0.5) \quad \text{or} \quad \rho_2 = 0.46$$

$$\rho_3 = \alpha_1\,\rho_2 + \alpha_2\rho_1 = 1.2 \times 0.46 - 0.5 \times 0.8 \quad \text{or} \quad \rho_3 = 0.152$$

$$\rho_4 = \alpha_1\rho_3 + \alpha_2\rho_2 = 1.2 \times 0.152 - 0.5 \times 0.46 \quad \text{or} \quad \rho_4 = -0.0476$$

$$\rho_5 = \alpha_1\rho_4 + \alpha_2\rho_3 = 1.2 \times (-0.0476) - 0.5 \times 0.152$$

$$\text{or} \quad \rho_5 = -0.13312$$

From Eq. 9.26, $$\sigma_x^2 = \frac{\sigma_\xi^2}{(1 - \alpha_1\rho_1 - \alpha_2\rho_2)}$$

$$\sigma_x^2 = \frac{1.0}{(1 - 1.2 \times 0.8 + 0.5 \times 0.46)} = \frac{1.0}{0.27} = 3.704.$$

9.5. AUTOREGRESSIVE MOVING AVERAGE PROCESS

A mixed autoregressive (of order p) and moving average (of order q) process is denoted as an $ARMA$ (p, q) process. Its general form is given by

$$(X_t - \mu_x) = \alpha_1\,(X_{t-1} - \mu_x) + \alpha_2\,(X_{t-2} - \mu_x) + \ldots\ldots + \alpha_p\,(X_{t-p} - \mu_x) + \xi_t - \beta_1\,\xi_{t-1} - \beta_2\,\xi_{t-2} - \ldots\ldots - \beta_q\,\xi_{t-q} \quad \ldots(9.42)$$

where ξ_t is an independent process with mean zero and $\alpha_1, \alpha_2, ...,$ α_p and $\beta_1, \beta_2, ..., \beta_q$ are the parameters of the process.

ARMA (1, 1) process. A mixed process of considerable practical importance is the *ARMA*(1, 1) process which is represented by

$$(X_t - \mu_x) = \alpha_1 (X_{t-1} - \mu_x) + \xi_t - \beta_1 \xi_{t-1} \qquad ...(9.43)$$

The variance of the process X_t and the variance of the independent process ξ_t are related by

$$\sigma_x^2 = \frac{(1 + \beta_1^2 - 2\alpha_1\beta_1)}{(1 - \alpha_1^2)} \cdot \sigma_\xi^2 \qquad ...(9.44)$$

The autocorrelation function is given by

$$\rho_1 = \frac{(1 - \alpha_1\beta_1)(\alpha_1 - \beta_1)}{(1 + \beta_1^2 - 2\alpha_1\beta_1)} \qquad ...(9.45)$$

$$\rho_k = \alpha_1 \rho_{k-1} ; \text{ for } k \geq 2 \qquad ...(9.46)$$

Thus, the autocorrelation function of *ARMA*(1, 1) process decays exponentially from a starting value of ρ_1 which depends on β_1 as well as α_1. If the ρ_1 and ρ_2 are known or given, α_1 may be obtained using Eq. (9.46) as $\alpha_1 = \rho_2/\rho_1$, and substituting the values of α_1 and ρ_1 in Eq. (9.45) provides solution for β_1. It can be shown that for *ARMA*(1, 1) to be stationary ρ_1 and ρ_2 must satisfy the following conditions.

$$\left.\begin{array}{l} |\rho_2| < |\rho_1| \\ \rho_2 > \rho_1 (2\rho_1 + 1) \text{ if } \rho_1 < 0 \\ \rho_2 > \rho_1(2\rho_1 - 1) \text{ if } \rho_1 > 0 \end{array}\right\} \qquad ...(9.47)$$

Example 9.6. *Obtain the first 4 serial correlation coefficients and the variance of an ARMA (1, 1) process represented by* $(X_t - \mu_x) = 0.75 (X_{t-1} - \mu_x) + \xi_t - 0.5 \xi_{t-1}$. *Assume* $\sigma_\xi^2 = 1.0$.

Sol. Comparing the given process with Eq. (9.43), we note that $\alpha_1 = 0.75$ and $\beta_1 = 0.5$. From Eq. (9.45)

$$\rho_1 = \frac{(1 - \alpha_1\beta_1)(\alpha_1 - \beta_1)}{(1 + \beta_1^2 - 2\alpha_1\beta_1)} = \frac{(1 - 0.75 \times 0.5)(0.75 - 0.5)}{1 + 0.5^2 - 2 \times 0.75 \times 0.5}$$

$$= \frac{0.625 \times 0.25}{1 + 0.25 - 0.75} = \frac{0.15625}{0.5} = 0.3125$$

$$\rho_2 = \alpha_1 \rho_1 = 0.75 \times 0.3125 = 0.2344$$

$$\rho_3 = \alpha_1\rho_2 = 0.75 \times 0.2344 = 0.1758$$

$$\rho_4 = \alpha_1\rho_3 = 0.75 \times 0.1758 = 0.1318$$

From Eq. (9.44),

$$\sigma_x^2 = \frac{\sigma_\xi^2 (1 + \beta_1^2 - 2\alpha_1\beta_1)}{(1 - \alpha_1^2)}$$

$$= \frac{1.0 \times (1 + 0.25 - 0.75)}{(1 - 0.5625)}$$

$$= \frac{0.5}{0.4375} = 1.1429.$$

9.6. AUTOREGRESSIVE INTEGRATED MOVING AVERAGE PROCESS

Let X_t be the given process. Then the first difference at time step t is given by $(X_t - X_{t-1})$. Therefore, the first difference series would be $(X_t - X_{t-1})$, $(X_{t+1} - X_t)$, $(X_{t+2} - X_{t+1})$, ... etc. Similarly the second difference series at time steps t, $t + 1$, $t + 2$, ... is given by $\{(X_t - X_{t-1}) - (X_{t-1} - X_{t-2})\}$, $\{(X_{t+1} - X_t) - (X_t - X_{t-1})\}$, $\{(X_{t+2} - X_{t+1}) - (X_{t+1} - X_t)\}$, ... etc. In a similar manner, the dth difference series can be written. In general the dth difference series at time step t is denoted by $\Delta^d X_t$. Let $W_t = \Delta^d X_t$. Let the series W_t be described by an $ARMA(p, q)$ process. That means

$$\Delta^d X_t = W_t = \alpha_1 W_{t-1} + \alpha_2 W_{t-2} + \ldots\ldots + \alpha_p W_{t-p} + \xi_t - \beta_1 \xi_{t-1} - \beta_2 \xi_{t-2} - \ldots\ldots - \beta_q \xi_{t-q} \quad \ldots(9.48)$$

Then the process X_t is called an *autoregressive integrated moving average process* of order (p, d, q) and abbreviated as *ARIMA* (p, d, q) process. The *ARIMA* models have been considered to model non-stationary time series. Since *ARMA* process is a stationary process, it means that the process X_t itself is not stationary but the dth difference of X_t is stationary. A special case of *ARIMA* (p, d, q) model is obviously the *ARMA* (p, q) model where no differencing is needed.

9.7. CORRELOGRAMS OF OTHER PROCESSES

If a series is complex periodic with a period T_p, the correlogram of this series is also a periodic function with the same period T_p. The amplitude of this correlogram is unity and the correlogram starts with $\rho_0 = 1$. The correlogram of a linear trend of the form $X_t = a + bt$ is a horizontal line with $\rho_k = 1$ for any k. In hydrology either complex periodic or a linear trend do not occur exclusively. They may occur only as a combination with other types of series. Therefore our interest shall be in the shapes of the correlograms of various composite series.

If a series is a simple combination of a periodic sine component and an independent stochastic component, its correlogram would also be periodic with same frequency as the sine component starting with a value of 1 at lag zero. However, its amplitude will not be unity as in the case of deterministic periodic series. But it will be less than unity. Similarly the correlogram of a series which is a combination of a complex periodic component and a stochastic component is also periodic whose frequency is equal to the fundamental frequency of the complex periodic component. But such a correlogram cannot reveal the other frequency components contained in the complex periodic series and their relative significance. This information can be obtained only through the spectral analysis which is discussed in Chapter 10.

Example 9.7. *The monthly run off series of a river for a period of 20 years (240 values) are analysed and the relevant statistical parameters are given in Table 9.1 and Table 9.2. Is the series stationary ? Why or why not ? What type of deterministic component is present in the series ?*

Table 9.1. Statistics of monthly run off of Example 9.7

Month	*Mean flow in tmcft*	*Standard deviation in tmcft*	*Correlation co-efficient with previous month*
June	8.81	8.57	0.1563
July	29.03	27.78	– 0.0378
Aug.	54.94	59.36	0.1778
Sept.	55.79	42.32	0.1802
Oct.	32.52	21.91	– 0.0388
Nov.	7.04	9.14	0.1005
Dec.	2.98	1.79	0.7869
Jan.	2.08	0.80	0.8485
Feb.	1.40	0.73	0.3395
March	137	1.17	0.8201
April	1.07	0.95	0.0591
May	0.49	0.26	0.5164

Table 9.2. The autocorrelation co-efficients of Example 9.7

Lag k in months	*Autocorrelation co-efficient r_k*	*k*	r_k	*k*	r_k
1	0.3981	13	0.3693	25	0.2909
2	0.1406	14	0.1034	26	0.1195
3	– 0.0811	15	– 0.0565	27	– 0.1202
4	– 0.1758	16	– 0.1944	28	– 0.2181
5	– 0.2304	17	– 0.2338	29	– 0.2513
6	– 0.2401	18	– 0.2463	30	– 0.2560
7	– 0.2356	19	– 0.2442	31	– 0.2556
8	– 0.1991	20	– 0.2044	32	– 0.2297
9	– 0.0831	21	– 0.0823	33	– 0.1096
10	0.1604	22	0.1688	34	0.1167
11	0.5157	23	0.2727	35	0.3464
12	0.4636	24	0.4997	36	0.4126

Sol. Let the given monthly series be arranged as an ensemble of 20 series each of length 12 months. The means, standard deviations and correlation coefficient with $\tau = 1$ month can be computed for different months. These are infact what are given in Table 9.1. As is evident from this table, the monthly run off series is non-stationary. This is because the mean, standard deviations and correlation coefficient are varying quite significantly from month to month exhibiting non-stationarity in these parameters, that is, in mean and covariance of lag one.

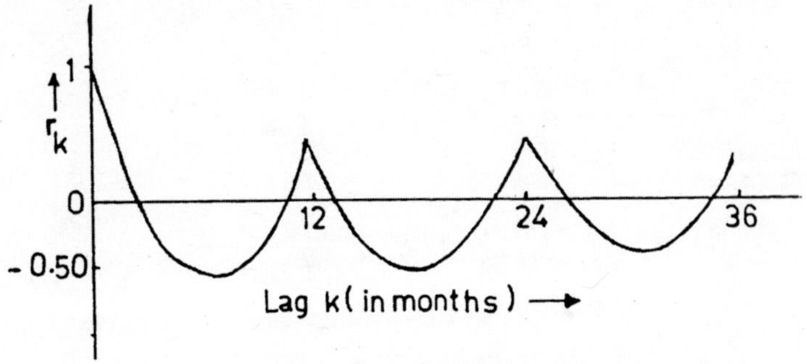

Fig. 9.6. Correlogram of monthly run off series of Example 9.7.

The computed autocorrelation coefficients upto 36 lags are plotted to give the correlogram of monthly run off series as shown in Fig. 9.6. This correlogram is fluctuating systematically with a period of 12 months or a frequency of 1 cycle/12 months. Therefore it can be concluded that the monthly run off series contains a deterministic periodic component with a fundamental frequency of 1 cycle/year.

Example 9.8. *Suppose that the following transformation is performed on the monthly run off series of Example 9.7.*

$$x'(i,j) = \frac{x(i,j) - \bar{x}(j)}{s(j)}$$

where $x(i, j)$ is the run off in jth month of ith year, $x'(i, j)$ is the corresponding transformed flow, with $j = 1, 2, \ldots\ldots, 12$ and $i = 1, 2, \ldots\ldots, 20$, $\bar{x}(j)$ and $s(j)$ are the mean and standard deviation of the run off in the jth month given by

$$\bar{x}(j) = \frac{1}{20} \sum_{i=1}^{20} x(i,j),\ j = 1, 2, \ldots\ldots, 12$$

$$s^2(j) = \frac{1}{20} \sum_{i=1}^{20} [x(i,j) - \bar{x}(j)]^2,\ j = 1, 2, \ldots\ldots, 12$$

Will the transformed flows constitute a second order stationary series ?

Sol. The given transformation cannot make the monthly run off series second order stationary. This is because though the transformed series will have monthly mean as 0 and the monthly variance as 1 for all the months, the correlation coefficient between any two successive months for the transformed series remains the same as that of the original series as explained below and since the correlation coefficient is not the same for all the months the transformed series is still non-stationary in covariance.

The correlation coefficient $r'(j)$ between the transformed flows of jth and $(j-1)$th months is given by

$$r'(j) = \frac{\frac{1}{20} \sum_{i=1}^{20} [x'(i,j) - 0]\,[x'(i,j-1) - 0]}{1 \times 1}$$

Since the mean and standard deviation of transformed flows are 0 and 1 respectively for all the months. Further from the definition of $x'(i, j)$

$$r'(j) = \frac{\frac{1}{20} \sum_{i=1}^{20} [x(i,j) - \bar{x}(j)]\,[x(i,j-1) - \bar{x}(j-1)]}{s(j)\, s(j-1)}$$

and this is equal to $r(j)$, the correlation coefficient between the flows of jth and $(j-1)$th in the original series.

The reader may note that while $\bar{x}(j)$, $s(j)$ and $r(j)$ are computed from ensemble of 20 series of each 12 months long, r_k are computed from the entire data as one sample function of 240 month long.

EXERCISES

9.1. The first 9 autocorrelation coefficients computed from 150 years of annual flow record at a river site are 0.461, 0.005, – 0.095, – 0.057, – 0.048, – 0.010, – 0.018, – 0.016 and – 0.055. Test the flow series for independence at 95% confidence level.

9.2. The $AR(1)$ model is written as $(X_t - \mu_x) = \alpha\,(X_{t-1} - \mu_x) + \xi_t$ where ξ_t is an independent process. If μ_x, σ_x^2 and ρ_1 are the mean, variance and first serial correlation coefficient of X_t, show that $\alpha = \rho_1$ and the mean and variance of ξ_t should be 0 and $\sigma_x^2\,(1 - \sigma_1^2)$ respectively.

9.3. Can you justify $AR(1)$ model with a negative ρ_1 to model the annual flows ?

9.4. A hydrologic variable is described by an $AR(1)$ model with mean, variance and first serial correlation coefficient of 100, 900 and 0.5 respectively. What is the standard deviation of the independent process ξ_t in the model.

9.5. Sketch the theoretical correlograms of an $AR(1)$ model with ρ_1 positive and also with ρ_1 negative, and also that of an independent process.

9.6. Verify whether the following processes are stationary :

(*i*) $(X_t - \mu_x) = 0.5\,(X_{t-1} - \mu_x) + \xi_t$

(*ii*) $(X_t - \mu_x) = \xi_t + 1.3\,\xi_{t-1} + 0.4\,\xi_{t-2}$

(*iii*) $(X_t - \mu_x) = 0.5\,(X_{t-1} - \mu_x) + \xi_t + 1.3\,\xi_{t-1} + 0.4\,\xi_{t-2}$

(*iv*) $(X_t - \mu_x) = 1.5\,(X_{t-1} - \mu_x) - 0.5\,(X_{t-2} - \mu_x) + \xi_t$

(*v*) $(X_t - \mu_x) = (X_{t-1} - \mu_x) + \xi_t + 0.5\,\xi_{t-1}$.

9.7. For all the five processes in Exercise 9.6 compute the first 5 serial correlation coefficients.

9.8. Show that for an $AR\,(m)$ process the autocorrelation coefficient of lag k is given by

$$\rho_k = \alpha_1\,\rho_{k-1} + \alpha_2\,\rho_{k-2} + \ldots\ldots + \alpha_m\,\rho_{k-m}.$$

where $\alpha_1, \alpha_2, \ldots\ldots, \alpha_m$ are the parameters of the process.

9.9. Write down Yule-Walker equations for an $AR\,(2)$ process and hence obtain the expressions for the two parameters of the process α_1 and α_2 in terms of ρ_1 and ρ_2.

9.10. A second order AR process is given by

$$(X_t - \mu_x) = (X_{t-1} - \mu_x) - 0.5\,(X_{t-2} - \mu_x) + \xi_t$$

(*i*) Calculate ρ_1 (*ii*) Using ρ_0 and ρ_1 as starting values calculate the values of ρ_k for $k = 2, 3, \ldots\ldots, 15$ from the result of Exercise 9.8 and plot the correlogram. From the plotted correlogram find the frequency of the damped sine wave. Compare this with the value given by Eq. (9.40).

9.11. Classify each of the processes in Exercise 9.6 as a member of *AR*, *MA* or *ARMA* processes.

9.12. The first serial correlation coefficient of *MA*(1) process is 0.3. Determine the parameter of the process.

9.13. A moving average process is given by $(X_t - \mu_x) = \xi_t - 1.2\,\xi_{t-1} + 0.5\,\xi_{t-2}$. Determine the variance of X_t, if the variance of ξ_t is 1.5.

9.14. If σ_ξ^2 is the variance of ξ_t in an *AR* (*m*) process, show that the variance of X_t is related to variance of ξ_t by

$$\sigma_\xi^2 = \sigma_x^2\,(1 - \alpha_1\,\rho_1 - \alpha_2\,\rho_2 - \ldots\ldots - \alpha_m\,\rho_m).$$

9.15. The first 10 serial correlation coefficients computed from 65 years of annual run off of a river are 0.49, 0.27, 0.10, 0.07, 0.02, 0.00, –0.004, – 0.002, 0.001, 0.00. Test the annual flow series for independence at 95% confidence level and suggest a suitable model to describe the series.

10

Spectral Analysis

10.1. INTRODUCTION

In the previous chapter we discussed the technique of autocorrelation analysis to describe a stochastic process. The correlogram discussed there is a function of time and therefore it is said to belong to time domain. In this chapter we shall present an equivalent method of describing a stochastic process which is provided by its spectrum. The spectrum of a process shows how the variance of the process is distributed over frequency. The term frequency used here is in the harmonic sense and not in the sense of histogram. The spectral methods therefore belong to the frequency domain.

10.2. LINE SPECTRUM

Consider a discrete series x_t defined at finite number of time points $t = 1, 2, \ldots\ldots, n$ which are spaced Δt apart. That is, the total lengths of the record $T = n.\Delta t$. Suppose that n is odd and equal to $(2q + 1)$. Therefore $q = (n - 1)/2$. Then this finite series of length T can be fitted by finite *Fourier series* consisting of trigonometric sine and cosine functions.

$$x_t = \bar{x} + \sum_{m=1}^{q} [a_m \cos(2\pi f_m t) + b_m \sin(2\pi f_m t)] \qquad \ldots(10.1)$$

where $f_m = m/n$ is the frequency of the mth *harmonic* and a_m and b_m are the coefficients of the mth harmonic. The frequency $f_1 = \frac{1}{n}$ cycles/Δt is called the *fundamental frequency* whose period is $n.\Delta t$ or T. Hence f_1 will be expressed in cycles/second if Δt is a second, in cycles/day if Δt is a day, in cycles/month if Δt is a month and so on. The least square estimates of a_m and b_m are given by

$$a_m = \frac{2}{n} \sum_{t=1}^{n} x_t \cos(2\pi f_m t) \text{ and } b_m = \frac{2}{n} \sum_{t=1}^{n} x_t \sin(2\pi f_m t)$$

Since $f_m = m/n$, we can write

$$a_m = \frac{2}{n} \sum_{t=1}^{n} x_t \cos\left(2\pi \frac{m}{n} t\right); m = 1, 2, ..., q \qquad ...(10.2)$$

$$b_m = \frac{2}{n} \sum_{t=1}^{n} x_t \sin\left(2\pi \frac{m}{n} t\right); m = 1, 2, ..., q \qquad ...(10.3)$$

Eq. (10.1) can also be written in the form

$$x_t = \bar{x} + \sum_{m=1}^{q} c_m \sin(2\pi f_m t + \phi_m) \qquad ...(10.4)$$

Where c_m and ϕ_m are respectively called the amplitude and the phase angle of the mth harmonic which are given by

$$c_m^2 = a_m^2 + b_m^2 \qquad ...(10.5)$$

$$\phi_m = \tan^{-1}(-b_m/a_m) \qquad ...(10.6)$$

A plot of $c_m^2/2$ *Versus* the frequency f_m is known as the *line spectrum*. It is also known as *discrete spectrum* or *variance spectrum* or *periodogram*. If s_x^2 is the variance of the series, then

$$s_x^2 = \sum_{m=1}^{q} (c_m^2/2) \qquad ...(10.7)$$

Thus we see that each harmonic with frequency f_m has a variance of $c_m^2/2$ and the sum of the variances of all the harmonics equals the variance of the series. The line spectrum consists of $q = (n - 1)/2$ values and displays the variance explained by each harmonic and incidentally the relative significance of the harmonics also.

When n is even, q is set equal to $n/2$ instead of $(n - 1)/2$ and Eqs. (10.2) and (10.3) are applied to obtain a_m and b_m for m = 1, 2, ..., $(q - 1)$ while a_q and b_q are given by

$$a_q = \frac{1}{n} \sum_{t=1}^{n} (-1)^t x_t \qquad ...(10.8)$$

$$b_q = 0 \qquad ...(10.9)$$

If in the Fourier series fitting of the series only a limited number of harmonics, instead of all q harmonics, are used then the fit will be only approximate and not exact. In such a case the harmonics which explain more than, say, 90 per cent of the observed variance may be included in the fitting.

Example 10.1. *The mean monthly run off of a river are computed from a 20 year record and are given below. Fit the Fourier series to this data and construct the line spectrum.*

Month	*Mean run off in* Mm^3
June	*249.5*
July	*822.9*
Aug.	*1557.3*
Sept.	*1581.4*
Oct.	*921.8*
Nov.	*199.5*
Dec.	*84.5*
Jan.	*59.0*
Feb.	*39.7*
Mar.	*38.8*
April	*30.3*
May	*13.9*

Sol. The mean $\bar{x}$ and the variance of the series are computed as follows.

$$\bar{x} = \frac{249.5 + 822.9 + \ldots\ldots + 13.9}{12} = 466.55$$

$$s_x^{\,2} = \frac{[(249.5 - 466.55)^2 + (822.9 - 466.55)^2 + \ldots\ldots + (13.9 - 466.55)^2]}{12}$$

$$= 329810$$

Since the record length $n = 12$ is even, the number of harmonics to be considered, q will be $n/2$, that is 6 and the coefficients a_m and b_m for these harmonics are to be computed from

$$a_m = \frac{2}{n} \sum_{t=1}^{n} x_t \cos\left(\frac{2\pi m t}{n}\right), \text{ for } m = 1, 2, 3, 4, 5$$

$$b_m = \frac{2}{n} \sum_{t=1}^{n} x_t \sin\left(\frac{2\pi m t}{n}\right), \text{ for } m = 1, 2, 3, 4, 5$$

$$a_6 = \frac{1}{n} \sum_{t=1}^{n} (-1)^t x_t \quad \text{and} \quad b_6 = 0.$$

The sample calculations, to obtain a_2 for example are as follows.

$$a_2 = \frac{2}{n} \sum_{t=1}^{n} x_t \cos\left(\frac{2\pi \cdot 2 \cdot t}{n}\right).$$ Since $n = 12$ in the present case,

we have $$a_2 = \frac{1}{6} \sum_{t=1}^{12} x_t \cos\left(\frac{\pi}{3} \cdot t\right)$$

$$a_2 = \frac{1}{6}[x_1 \cos(1 \times 60°) + x_2 \cos(2 \times 60°) + x_3 \cos(3 \times 60°) + \ldots\ldots + x_{12} \cos(12 \times 60°)]$$

$$= \frac{1}{6}[x_1 \cos 60° + x_2 \cos 120° + x_3 \cos 180° + \ldots\ldots + x_{12} \cos 720°]$$

$$= \frac{1}{6}[249.5 \times \cos 60° + 822.9 \times \cos 120° + \ldots\ldots + 13.9 \times \cos 720°]$$

$$= \frac{1}{6}[124.75 - 411.45 - 1557.3 - \ldots\ldots + 13.9]$$

$$= -331.93.$$

Similarly all other coefficients are calculated and are given in Table 10.1.

Table 10.1. Fourier decomposition of data of Example 10.1

Harmonic	*m*	a_m	b_m	$c_m = \sqrt{a_m^2 + b_m^2}$	ϕ_m	*Contribution to variance* $c_m^2/2$	*% contribution to variance*
Fundamental	1	– 200.7	673.9	703.2	73°.40	247211	74.96
Second	2	– 331.9	– 195.8	385.4	– 30°.54	74248	22.51
Third	3	98.9	– 76.9	125.3	37°.87	7847	2.38
Fourth	4	– 14.0	17.5	22.4	51°.34	251	0.076
Fifth	5	9.0	8.1	12.1	41°.99	73	0.022
Sixth	6	– 14.0	0	14.0	0°.00	98	0.030
					Total	329728	99.978

It is seen from this table that the total of the variances contributed by all the six harmonics is equal to the variance in the series within rounding off errors. It is also seen that the fundamental and the second harmonic account for more than 97% (74.96 + 22.51 = 97.47) of the variance in the series. Therefore, using Eq. (10.1) with $q = 2$, the data series may be well approximated by

$$\hat{x}_t = 466.55 - 200.7 \cos\left(\frac{2\pi t}{12}\right) + 673.9 \sin\left(\frac{2\pi t}{12}\right) - 331.9 \cos\left(\frac{4\pi t}{12}\right) - 195.8 \sin\left(\frac{4\pi t}{12}\right)$$

or $$\hat{x}_t = 466.55 - 200.7 \cos(\pi t/6) + 673.9 \sin(\pi t/6) - 331.9 \cos(\pi t/3) - 195.8 \sin(\pi t/3).$$

Alternatively using Eq. (10.4), we can also write

$$\hat{x}_t = 466.55 + 703.2 \cos\left(\frac{\pi t}{6} + 73°.4\right) + 385.4 \cos\left(\frac{\pi t}{3} - 30°.54\right).$$

The periodogram constructed from the results of Table 10.1 is shown in Fig. 10.1 which depicts the contribution to the variance by each harmonic.

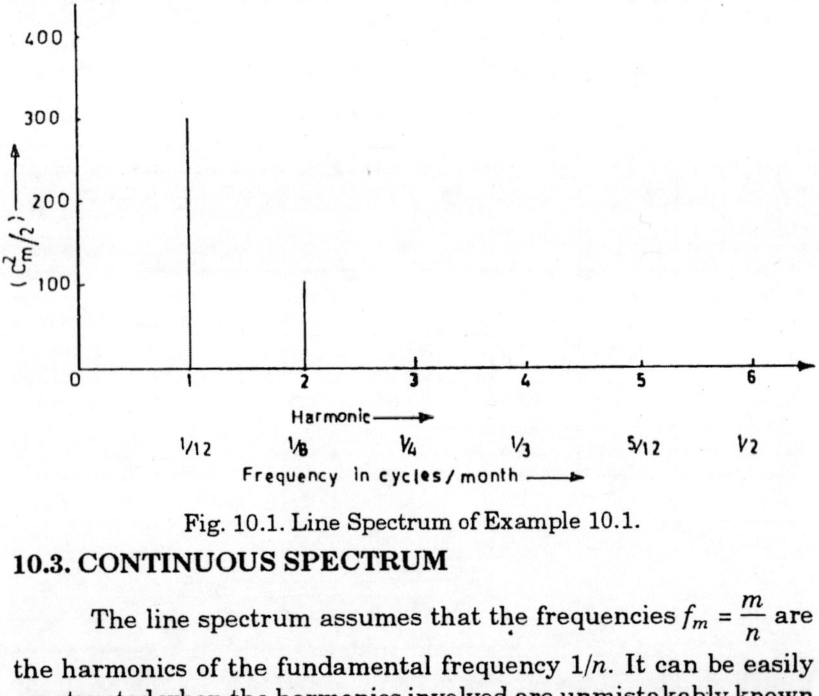

Fig. 10.1. Line Spectrum of Example 10.1.

10.3. CONTINUOUS SPECTRUM

The line spectrum assumes that the frequencies $f_m = \frac{m}{n}$ are the harmonics of the fundamental frequency $1/n$. It can be easily constructed when the harmonics involved are unmistakably known and are small in number. When the series is of large length the number of harmonics to describe the data by Fourier series will also be large resulting is lengthy and tedious calculations. Similarly when the series is not discrete but continuous we require theoretically infinite number of harmonics to fit the data exactly. To overcome the difficulties in the computation of line spectrum in situations where the basic harmonics are not known, and the frequency f_m is not related to the length of the series, the continuous spectrum of the series is constructed. Let $\psi(f)$ be the ordinate of the continuous spectrum corresponding to any frequency f. Then, we define $\psi(f)$ in such a way that, the variance contributed by all the frequencies in the interval Δf around f is given by $\psi(f) \cdot \Delta f$. Thus we see that $\psi(f)$ does not indicate the variance contributed by the frequency f as such, but it does indicate the concentration or density of the variance disposition at f. The difference, then, between line spectrum and continuous spectrum is similar to the difference between the probability mass function of a discrete random variable and the probability density function of a continuous random variable. Therefore the continuous spectrum may also be appropriately called as *variance density spectrum*. For any process, the continuous spectrum has to satisfy the following two conditions :

(*i*) Since variance cannot be negative, $\gamma(f) \geq 0$.

(*ii*) Since the total area under continuous spectrum shall represent the variance of the process, $\int_0^\infty \psi(f)\, df = \sigma_x^2$.

It is sometimes necessary to compare different time series which have different scales of measurement. In such circumstances, it is convenient to normalise $\psi(f)$ by dividing it by the variance of the series σ_x^2. Accordingly, the function $\gamma(f)$ given by $\psi(f)/\sigma_x^2$ is called the *spectral density function*. Evidently the spectral density function has to satisfy the condition that $\int_0^\infty \gamma(f) df = 1$.

Fortunately the spectral density function can be computed from the correlogram itself. For a discrete time series the spectral density function $\gamma(f)$ at any frequency f is given by

$$\gamma(f) = 2\left[1 + 2 \sum_{k=1}^{\infty} \rho_k \cos(2\pi f k)\right] \qquad \text{...(10.10)}$$

where ρ_k is the population autocorrelation coefficient of lag k.

In practice, however, the values of $\gamma(f)$ are estimated by the *sample spectral density function* $g(f)$ using the sample autocorrelation coefficient r_k in place of ρ_k. Thus $g(f)$ is given by

$$g(f) = 2\,[1 + 2 \sum_{k=1}^{m} r_k \cos(2\pi\, fk)] \qquad \text{...(10.11)}$$

where m is the maximum lag upto which the serial correlation coefficients are estimated. The spectral density function is usually estimated for the frequency range $0 \leq f \leq 0.5$. The frequency of 0.5

cycles/unit time is called the *Nyquist frequency* and it is the highest frequency which can be detected from data sampled at intervals of unit time. Thus if one is analysising daily series, maximum frequency is 0.5 cycles/day and for monthly data it is 0.5 cycles/month and so on.

10.4. SMOOTHING OF SPECTRAL DENSITY ESTIMATES

It has been shown by many investigators that the spectral density estimates obtained from Eq. (10.11) are biased and not efficient. It may be shown that the sample spectrum of a stationary time series fluctuates violently about the theoretical spectrum. An intuitive explanation for this is that the sample spectrum corresponds to using an interval Δf in the frequency domain whose width is too small. This is analogous to using too small a group interval for the histogram when estimating an ordinary probability density.

To obtain unbiased and efficient estimates, a modification to Eq. (10.11) is necessary. This modification is similar to using $(n-1)$ instead of n in the denominator in the estimation of sample variance. The modification may be in the lag domain (or time domain) on the autocorrelation function or in the frequency domain on $g(f)$. This procedure is called the *smoothing of spectral density function.* For this reason, the spectral density estimates given by Eq. (10.11) is called the *raw spectral density*.

When smoothing is effected in the lag domain, that is on the autocorrelation function, the expression for the smooth spectral density $s(f)$ is given by

$$s(f) = 2\left[1 + 2\sum_{k=1}^{m} D(k)\, r_k \cos(2\pi f k)\right] \qquad \text{...(10.12)}$$

where $D(k)$ is variously known as *smoothing function in the lag domain* or *smoothing window or weighting function.* There are many smoothing windows suggested in the literature. A few of them are mentioned below.

Bartlett's Window

$$\begin{aligned} D(k) &= 1 - \frac{k}{m} && \text{for } k \le m \\ &= 0 && \text{for } k > m \end{aligned} \qquad \text{...(10.13)}$$

Parzen's Window

$$\text{D}(k) = 1 - 6\left(\frac{k}{m}\right)^2 + 6\left(\frac{k}{m}\right)^3 \quad \text{for } k \le \frac{m}{2}$$

$$= 2\left(1 - \frac{k}{m}\right)^3 \quad \text{for } \frac{m}{2} < k \le m$$

$$= 0 \quad \text{for } k > m \quad \text{...(10.14)}$$

Blackman and Tukey (Hanning) Window

$$D(k) = \frac{1 + \cos(\pi k/m)}{2} \quad \text{for } k \le m$$

$$= 0 \quad \text{for } k > m \quad \text{...(10.15)}$$

The most commonly used smoothing window in hydrology is the one given by Eq. (10.15).

It is recommended that the values of the spectral density function $s(f)$ be calculated only at $(m + 1)$ special discrete frequencies f_k given by

$$f_k = \frac{k \cdot f_c}{m}$$

$$k = 0, 1, 2, \ldots\ldots, m \quad \text{...(10.16)}$$

where f_c is the Nyquist frequency. The index k is called the harmonic number. Thus if the correlogram is estimated for 10 lags, the spectral density function is estimated at 11 frequencies, and if the correlogram is estimated for 36 lags, the spectral density is estimated at 37 frequencies and so on and so forth.

An equivalent smoothing procedure in the frequency domain which corresponds to the Blackman and Tukey window is given by

$$\left.\begin{aligned} s(f_k) &= \frac{1}{4} g(f_{k-1}) + \frac{1}{2} g(f_k) + \frac{1}{4} g(f_{k+1}) \\ &\text{for } 0 < k < m \\ s(f_0) &= \frac{1}{2} g(f_0) + \frac{1}{2} g(f_1) \\ s(f_m) &= \frac{1}{2} g(f_{m-1}) + \frac{1}{2} g(f_m) \end{aligned}\right\} \quad \text{...(10.17)}$$

where f_0 and f_m correspond to the zero and Nyquist frequencies respectively.

Alternative to Eq. (10.17), the hamming method due to R.W. Hamming which takes the form given in Eq. (10.17 *a*) could be used.

$$\left.\begin{aligned} s(f_k) &= 0.23\, g(f_{k-1}) + 0.54\, g(f_k) + 0.23\, g(f_{k+1}) \\ &\text{for } 0 < k < m \\ s(f_0) &= 0.54\, g(f_0) + 0.46\, g(f_1) \\ s(f_m) &= 0.54\, g(f_{m-1}) + 0.46\, g(f_m) \end{aligned}\right\} \quad \text{...(10.17 a)}$$

The concept of continuous spectrum is of recent origin compared to the line spectrum. Though the later developments have overshadowed the application of discrete spectra and made them appear inferior as an investigative technique, it may be still advantageous to use the line spectrum in hydrology whenever the basic periods are unmistakenly known from the physical considerations. When the basic frequency and amplitudes of the harmonics are not known in advance and must be estimated, the use of the continuous spectrum is recommended.

10.5. SPECTRAL DENSITY FUNCTION OF INDEPENDENT PROCESS

In an independent process or a pure random process, no individual frequency is more predominant than any other frequency. Accordingly the theoretical spectral density function of a pure random process should have a constant value over the frequency range $0 \leq f \leq f_c$ as shown in Fig. 10.2. Since the total area under spectral density function must equal unity, this constant ordinate should be equal to 2.

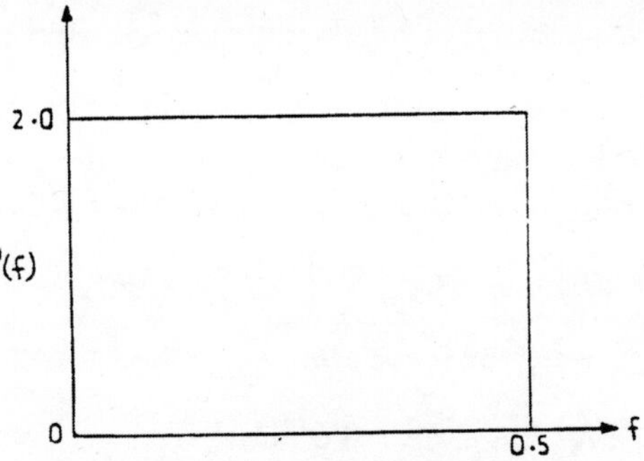

Fig. 10.2. Spectral density function of an independent process.

The spectral analysis finds a physical analogy in the prismatic resolution of a beam of sunlight. Hence the name spectral analysis. Fig. 10.3 brings out this analogy schematically. When a white light is passed through a prism it results in a decomposition consisting of spectrum of equal intensity. If the source of light is an artificial incandescent element, the spectrum consists of a finite number of bright lines of unequal intensity.

The spectrum of a white light and the spectral density function of an independent process are identical with constant value at all frequencies. Based on this analogy an independent process is also called a *white noise* process.

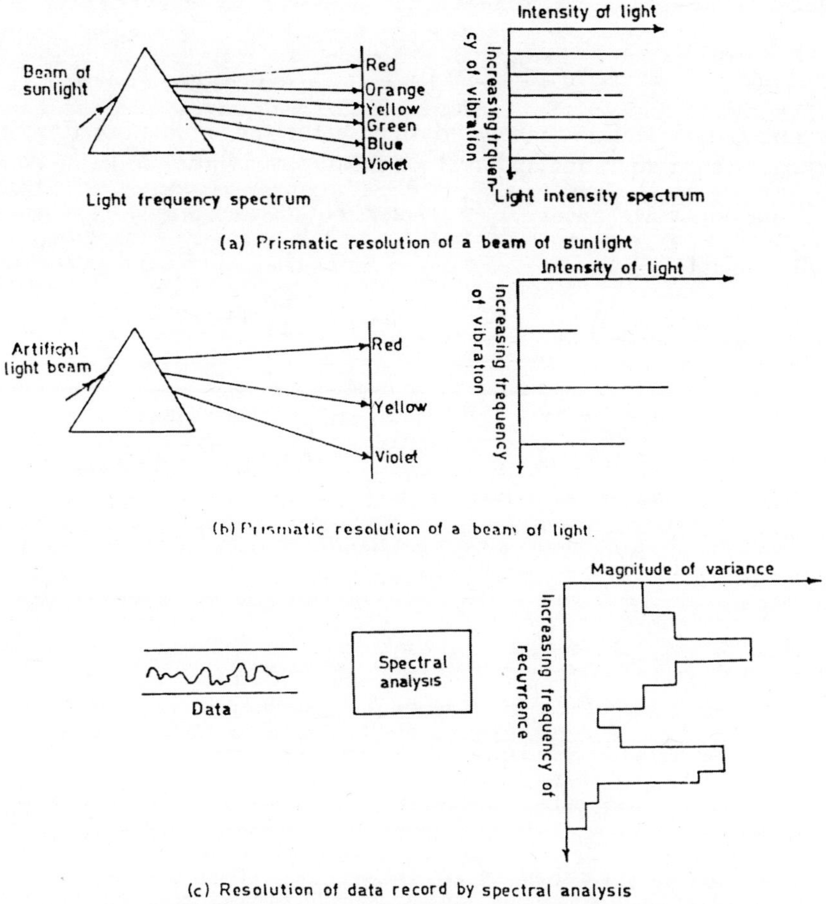

Fig. 10.3. Analogy between spectral analysis and prismatic resolution of light.

Though, theoretically the spectral density of a pure random process has to be constant, when it is estimated from a sample function, it will not be exactly equal to 2, and it may be fluctuating around this value. Therefore we have to ascertain whether these fluctuations are only due to sampling and well with in the acceptable limits or whether they are reflections of other components present in the series. It has been shown that the quantity $\left[\frac{\nu \cdot s(f)}{\gamma(f)} \right]$ follows a chi-square distribution with ν degrees of freedom, where ν depends on the type of window used for smoothing of the spectral density function. For the Blackman and Tukey window the number of degrees of freedom is given by

$$v = \frac{2n}{m} - \frac{2}{3} \qquad \ldots(10.18)$$

where n in the total number of data points in the sample function, and m is the maximum lag for which the autocorrelation function is estimated. Now, since $\frac{v \,.\, s(f)}{\gamma(f)}$ is distributed according to χ_v^2, it follows that

$$P\left[\chi_v^2\left(\frac{\alpha}{2}\right) \le \frac{v \,.\, s(f)}{\gamma(f)} \le \chi_v^2\left(1 - \frac{\alpha}{2}\right)\right] = 1 - \alpha$$

where the values of $\chi_v^2\left(\frac{\alpha}{2}\right)$ and $\chi_v^2\left(1 - \frac{\alpha}{2}\right)$ are the values obtained from the cumulative distribution function of chi-square distribution with v degrees of freedom such that

$$P\left[\chi_v^2 \le \chi_v^2\left(\frac{\alpha}{2}\right)\right] = \frac{\alpha}{2}$$

and
$$P\left[\chi_v^2 \le \chi_v^2\left(1 - \frac{\alpha}{2}\right)\right] = 1 - \frac{\alpha}{2}$$

Therefore if $(1 - \alpha)$ 100 is the stipulated confidence level, then the confidence interval is given by

$$\chi_v^2\left(\frac{\alpha}{2}\right) \le \frac{v \,.\, s(f)}{\gamma(f)} \le \chi_v^2\left(1 - \frac{\alpha}{2}\right)$$

$$\frac{\chi_v^2\left(\frac{\alpha}{2}\right)}{v \,.\, s(f)} \le \frac{1}{\gamma(f)} \le \frac{\chi_v^2\left(1 - \frac{\alpha}{2}\right)}{v \,.\, s(f)}$$

$$\frac{v \,.\, s(f)}{\chi_v^2\left(1 - \frac{\alpha}{2}\right)} \le \gamma(f) \le \frac{v \,.\, s(f)}{\chi_v^2(\alpha)} \qquad \ldots(10.19)$$

The confidence limits can now be fixed using Eq. (10.19) substituting the average value of the estimated spectral density $\bar{s}(f)$ in place of $s(f)$. If the estimated spectral density function lies completely within these confidence limits at all frequencies, the series may be inferred as pure random. If the result of the test is contrary to this, further investigations are to be carried out by way of comparing the estimated spectral density function with those of some standard processes. The following example will clearly demonstrate the procedure of testing the spectral density for independence.

Example 10.2. *Utilising the autocorrelation function of the monthly run off series of 20 years of Example 9.7, the smoothed spectral density function employing the Blackman and Tukey window are estimated from Eq. (10.12) and are given below. Is the monthly run off series independent ? If not what components are present in the data ?*

Harmonic	*Smoothed spectra s(f)*	*k*	*s(f)*	*k*	*s(f)*
0	2.85				
1	2.32	13	3.46	25	0.84
2	1.18	14	1.63	26	1.18
3	0.63	15	1.00	27	1.36
4	0.89	16	0.59	28	0.88
5	7.43	17	0.87	29	0.64
6	13.97	18	1.36	30	0.71
7	7.88	19	1.56	31	0.58
8	1.23	20	1.55	32	1.14
9	0.73	21	1.17	33	1.75
10	0.58	22	0.67	34	1.15
11	2.45	23	0.65	35	0.64
12	4.70	24	0.81	36	0.74

Sol. The given spectral density function is plotted in Fig. 10.4. The horizontal axis indicates both harmonic and the corresponding frequency in cycles/month. The spectrum is showing significant spikes at 6th and 12th harmonics corresponding to frequencies of 1 cycle/year and 2 cycles/year, indicating that the monthly series is not pure random. However, the confidence limits at 95% level are computed as given below :

The number of years of record

$$= 20$$

The number of monthly data points

$$n = 20 \times 12 = 240$$

Max. number of lags for which r_k is computed

$$m = 36$$

The number of degrees of freedom for the Chi-square distribution

$$\nu = \frac{2n}{m} - \frac{2}{3}$$

$$= \frac{2 \times 240}{36} - \frac{2}{3}$$

$$= 12.667 \simeq 13$$

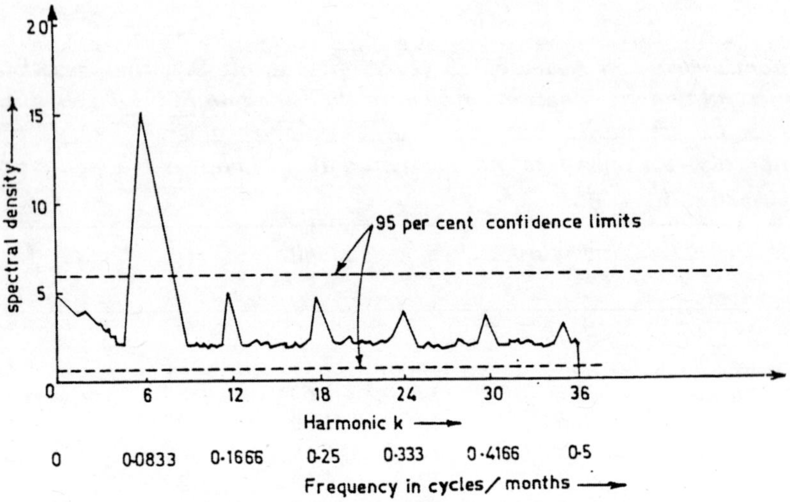

Fig. 10.4. Spectral density function of Example 10.2.

$$\text{Confidence level} = 95\% = (1 - \alpha)\,100$$

$$\therefore \qquad \alpha = 0.05$$

and
$$\frac{\alpha}{2} = 0.025$$

$$1 - \frac{\alpha}{2} = 0.975$$

For 13 degrees of freedom, from, the Chi-square distribution of Table A.2 in the Appendix.

$$\chi_{13}^2\,(0.025) = 5.01$$

and
$$\chi_{13}^2\,(0.975) = 24.74$$

The average spectral density $\bar{s}(f)$ is computed as

$$\bar{s}\,(f) = \frac{1}{37}\,(2.85 + 2.32 + 1.18 + \ldots.. + 1.15 + 0.64 + 0.74)$$

$$= \frac{73.77}{37} = 1.994$$

From Eq. 10.19, we have

$$\frac{13 \times 1.994}{24.74} < \gamma\,(f) \le \frac{13 \times 1.994}{5.01}$$

$$1.048 < \gamma\,(f) \le 5.173$$

These confidence limits are also shown in Fig. 10.4. Since the spectrum is not lying entirely within the confidence interval, the monthly series is not a pure random series. The high value of spectral density corresponding to the frequency of 1 cycle/year indicates that there is a cyclic component present in the series.

10.6. SPECTRUM OF A MOVING AVERAGE PROCESS

The theoretical spectral density of a *MA* (*m*) process represented by Eq. (9.15) is given by

$$\gamma(f) = 2 \left| 1 + \beta_1 e^{-j\,2\pi\,.\,1f} + \beta_2 e^{-j\,2\pi\,.\,2f} + \ldots\ldots + \beta_m e^{-j\,2\pi\,.\,mf} \right|^2 \qquad \ldots(10.20)$$

for $0 \le f \le 0.5$, where $j = \sqrt{-1}$.

The spectrum of *MA* (1) process can be obtained from Eq. (10.20) by setting $m = 1$. That is, for *MA* (1) process

$$\gamma(f) = 2 \left| 1 + \beta_1 e^{-j2\pi f} \right|^2$$

or

$$\gamma(f) = 2(1 + \beta_1^2 + 2\beta_1 \cos 2\pi f) \qquad \ldots(10.21)$$

Similarly the spectrum of *MA* (2) process is given by

$$\gamma(f) = 2 \left[1 + \beta_1^2 + \beta_2^2 + 2\beta_1 (1 + \beta_2) \cos 2\pi f + 2\beta_2 \cos 4\pi f \right] \qquad \ldots(10.22)$$

for $0 \le f \le 0.5$.

10.7. SPECTRUM OF AN AUTOREGRESSIVE PROCESS

The theoretical spectrum of an *AR*(*m*) process represented by Eq. (9.20) is given by

$$\gamma(f) = \frac{2}{\left| 1 - \alpha_1 e^{-j\,2\pi\,.\,1f} - \alpha_2 e^{-j\,2\pi\,.\,2f} - \ldots\ldots - \alpha_m e^{-j\,2\pi\,.\,mf} \right|^2} \qquad \ldots(10.23)$$

for $0 \le f \le 0.5$, where $j = \sqrt{-1}$.

AR(1) process. The spectrum of *AR*(1) process can be obtained by letting $m = 1$ in Eq. (10.23). That is, for *AR*(1) process

$$\gamma(f) = \frac{2}{\left| 1 - \alpha_1 e^{-j\,2\pi f} \right|^2}$$

or

$$\gamma(f) = 2/(1 + \alpha_1^2 - 2\alpha_1 \cos 2\pi f) \qquad \ldots(10.24)$$

In the *AR*(1) process, since $\alpha_1 = \rho_1$, the theoretical spectrum can be constructed by computing $\gamma(f)$ for various values of f in the range $0 \le f \le 0.5$ by replacing α_1 with ρ_1. The spectrum for an *AR*(1) process, both with ρ_1 positive and ρ_1 negative, is shown in Fig. 10.5. While the spectrum for the case of ρ_1 positive is dominated by low frequencies, it is higher frequencies which dominate the spectrum with ρ_1 negative. It may also be noted that $\gamma(f)$ for ρ_1 positive is equal to $\gamma(0.5 - f)$ for ρ_1 negative.

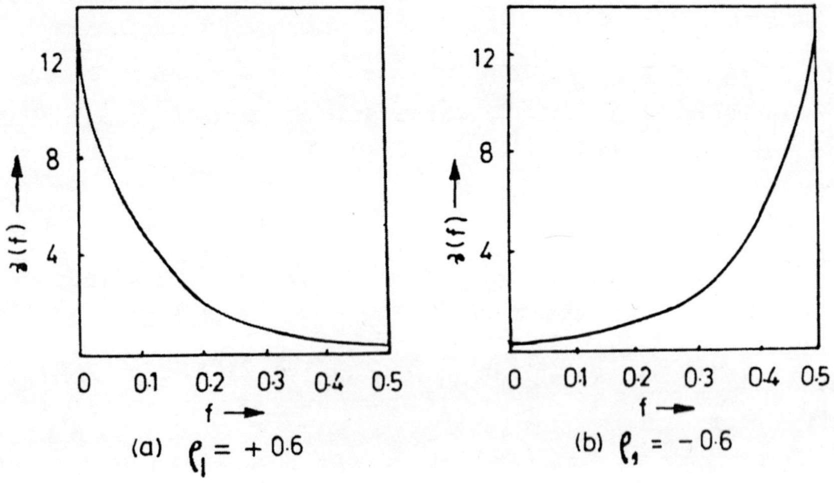

Fig. 10.5. Spectrum of $AR(1)$ process.

AR(2) process. The spectrum of $AR(2)$ process is obtained by substituting 2 for m in Eq. (10.23).

$$\gamma(f) = \frac{2}{\left[1 + \alpha_1^2 + \alpha_2^2 - 2\alpha_1 (1 - \alpha_2) \cos 2\pi f - 2\alpha_2 \cos 4\pi f\right]} \quad ...(10.25)$$

for $0 \le f \le 0.5$. As mentioned in Section 9.4, when $\alpha_1^2 + 4\alpha_2 < 0$ the correlogram will exhibit a damped sine wave form. The spectrum also reflects this pseudo-periodic behaviour for this case showing that a large portion of the variance of the series is accounted for by frequencies in the neighbourhood of f_0 given by Eq. (9.40). For example, if we consider an $AR(2)$ process with $\alpha_1 = 1$ and $\alpha_2 = -0.5$, then $(\alpha_1^2 + 4\alpha_2)$ will be less than zero and f_0 as per Eq. (9.40) is 0.125. The spectrum computed for this case is shown in Fig. 10.6 which has a peak ordinate at this frequency of 0.125.

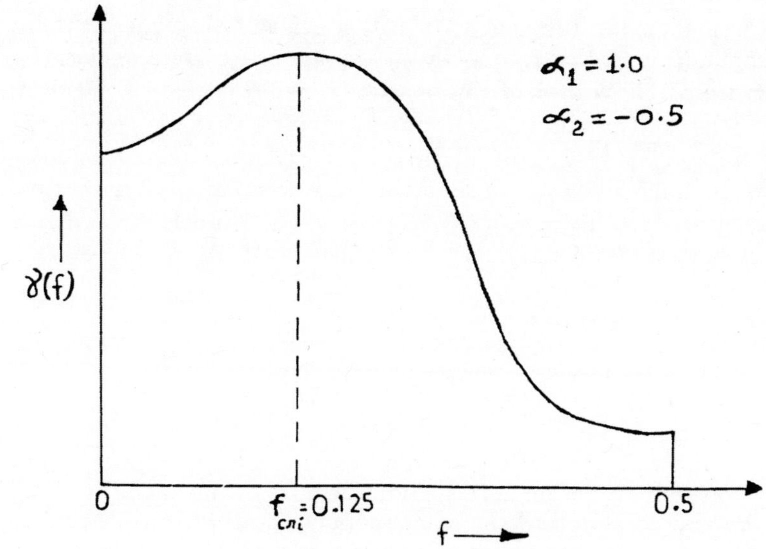

Fig. 10.6. Spectrum of *AR*(2) process.

10.8. SPECTRUM OF AUTOREGRESSIVE-MOVING AVERAGE PROCESS

The spectrum of an *ARMA* (*p*, *q*) process represented by Eq. (9.42) is given by

$$\gamma(f) = \frac{2\left|1 + \beta_1 e^{-j\,2\pi\,.\,1f} + \beta_2 e^{-j\,2\pi\,.\,2f} + \ldots\ldots + \beta_q e^{-j\,2\pi\,.\,qf}\right|^2}{\left|1 - \alpha_1 e^{-j\,2\pi\,.\,1f} - \alpha_2 e^{-j\,2\pi\,.\,2f} - \ldots\ldots - \alpha_p e^{-j\,2\pi\,.\,pf}\right|^2} \qquad \ldots(10.26)$$

for $0 \le f \le 0.5$, where $j = \sqrt{-1}$. This is also known as a *rational* spectrum. Specifically the spectrum of an *ARMA*(1, 1) process which is of interest to hydrologists is given by

$$\gamma(f) = 2\left|1 + \beta_1 e^{-j\,2\pi f}\right|^2 \Big/ \left(\left|1 - \alpha_1 e^{-j\,2\pi f}\right|^2\right)$$

$$\gamma(f) = \frac{2\,(1 + \beta_1^2 + 2\beta_1 \cos 2\pi f)}{(1 + \alpha_1^2 - 2\alpha_1 \cos 2\pi f)},\ 0 \le f \le 0.5. \qquad \ldots(10.27)$$

10.9. SPECTRUMS OF OTHER PROCESSES

If the discrete series is a periodic series made up of a simple sine or cosine function with frequency f_0, such as the one given by

Eq. (8.7), then its line spectrum will consist of only one ordinate corresponding to the frequency f_0. At all other frequencies the spectrum will have zero ordinate.

The continuous spectrum of a complex periodic series will have spikes of large magnitudes at all those frequencies corresponding to the harmonics which constitute the complex periodic series. These spikes are represented by the *dirac-delta* functions. A dirac-delta function at any frequency f_1 is mathematically described as

$$\left.\begin{aligned} &\delta(f) = 0 && \text{for} \quad f \neq f_1 \\ &\delta(f) \rightarrow \infty && \text{when } f = f_1 \\ &\int_{f_1-\varepsilon}^{f_1+\varepsilon} \delta(f)df = 1 && \text{for any } \varepsilon > 0 \end{aligned}\right\} \quad \text{...(10.28)}$$

The spectrum of a linear trend will have a spike represented by a dirac-delta function at the origin. This is because a linear trend may be viewed as a periodic component with an infinite period or zero frequency.

Deterministic periodic processes or deterministic linear trends are seldom observed in hydrology in isolation. As already pointed out the hydrologic series will usually be in the form of a combination of both deterministic and stochastic components. If the deterministic component of the hydrologic series is a complex periodic, then its spectrum will be dominated by the presence of spikes at the corresponding frequencies. The period represented by the first frequency will be the fundamental period of the series. The other frequencies are called the subharmonics. There may not be any physical significance to all these frequencies. Nevertheless, it means that these frequencies are required in modelling the deterministic component which is complex periodic. It may be noted here that while autocorrelation analysis reveals only the fundamental period of the deterministic component, the spectral analysis reveals not only the fundamental period but also the subharmonics which are significant and required in describing the deterministic periodic component.

Any hydrologic series which is observed seasonally with a time interval less than a year (like daily, weekly, and monthly series) is bound to have a cyclic component with a fundamental period of one cycle/year. This periodicity may be present in mean, variance and covariance. Now if the periodicity is removed from the series by fitting an appropriate deterministic periodic function the resulting residual series is free from cyclicity and may be assumed to be second order stationary. It is quite easy to model the periodicity in mean and variance. But the same is not true with covariance. Therefore, the common practice is to remove the periodicity in mean and variance only and to assume the residual series to be second order stationary. But then the residual series need not necessarily be stationary and quite often the discrepancy is considerable. This difficulty can be overcome to some extent by resorting to some transformation like logarithmic, square root or nth root. The original series which is nonstationary in covariance may become stationary under one of these transformations. If the periodicity in mean and variance is now removed from the transformed series, the residuals may be assumed to be second order stationary. The experience suggests that the logarithmic transformation is quite adequate in many cases. Further analysis on the residual series will result in the appropriate model for the stochastic component. These procedures are explained in detail in the next chapter.

EXERCISES

10.1. Sketch the theoretical spectrums of an $AR(1)$ process for the cases of (*i*) $\alpha_1 = 0.5$ (*ii*) $\alpha_1 = -0.5$ (*iii*) $\alpha_1 = 0.99$ and comment.

10.2. Construct the theoretical spectrum of an *ARMA* (1, 1) process for the cases of (*i*) $\alpha_1 = 0.3$ and $\beta_1 = 0.9$ and (*ii*) $\alpha_1 = 0.3$ and $\beta_1 = -0.9$.

10.3. Plot the spectrum of Exercises 9.10 and show that it has a peak at the frequency given by Eq. (9.40).

10.4. Explain how the spectrum of a series is tested for independence.

10.5. What is meant by smoothing of spectral density function ? Explain different methods of smoothing in lag domain and frequency domains.

10.6. Sketch the theoretical spectrum of a white noise process.

10.7. What is the effect of smoothing the spectral density function by using a window ? Write an expression for the Blackman and Tukey window in lag domain and its equivalent procedure in frequency domain.

10.8. Distinguish between the line spectrum and continuous spectrum. When is each preferred to the other ?

10.9. Explain how the ordinates of spectral density function are computed from the autocorrelation function.

10.10. The monthly run off data of 40 years is used to compute the first 36 autocorrelation coefficients. The ordinates of the spectral density function estimated using these autocorrelation coefficients at the harmonics 0, 1, 2, ..., 36 are as given below.

32.36, 13.86, 0.13, 0.15, 0.12, 4.93, 11.33, 4.80, 0.04, 0.05, 0.04, 1.08, 2.47, 1.07, 0.05, 0.04, 0.02, 0.19, 0.41, 0.21, 0.05, 0.04, 0.03, 0.04, 0.08, 0.06, 0.05, 0.04, 0.03, 0.02, 0.03, 0.03, 0.04, 0.03, 0.03, 0.02, 0.01

Test the series for independence at 95% tolerance level. What components of the series are present in the data ?

11

Univariate Stochastic Models

11.1. INTRODUCTION

The planning and management of water resource systems require large hydrological data over space and time. The data used in the design, planning and operational studies of water resource systems could be restricted to historical records. However, the serious drawback may then result because a particular sequence of observations does not occur in an identical form over a future period. Also available hydrologic data base is seldom large enough. But it is often difficult to justify delays in decision making pending the aquisition of additional data because the resulting benefits would have to be foregone. A feasible course of action is therefore to develop and use mathematical models to generate the data at specific points in space and time. Generated data sets are particularly useful in determining the reliabilities of performance of the complex systems.

The concept and application of stochastic models is of course based on the following assumptions. First, it is assumed that the sample data represents the population to a sufficient degree. When justified the conditions of stationarity and ergodicity are invoked and stationary models developed. Otherwise appropriate non-stationary models can be formulated. Secondly it is assumed that the errors of measurement and recording which exist in the data can be identified and rectified in the preliminary analysis. Also any man-made effects on the time series caused by river diversions, construction of dams etc., must be accounted for so that the data series after these adjustments represents natural events as they occur. Lastly the model, on the one hand, should be manageable in form, and on the other hand, it has to preserve the pertinent statistical and hydrological characteristics of the process so that various types of events and sequences that are likely to occur are generated in a realistic manner. In paractice, these conditions may not be fully met. Nevertheless, the best possible use must be made of the available data.

The essential procedure in model building is that after identifying the correct type of model for describing the process, its parameters are to be properly estimated and adjusted, if necessary, so that the important stochastic structure noticed in the observed sample function is preserved in the generation model. In other words, if the model is used to generate a synthetic sequence, such a sequence should be statistically indistinguishable from its historical counterpart. More specifically if one estimates the statistical parameters like mean, standard deviation, serial correlation coefficient etc., of a synthetic sequence they should be identical, in a statistical sense, with the corresponding parameters estimated from a historical sequence. Apart from preserving the important statistical parameters like mean, standard deviation and correlation structure the model should also preserve the probability distribution of the observed flows. That means, if the observed flows follow a normal distribution, the generated flows should also follow normal distribution, and if the observed flows follow log-normal distribution the generated flows should also follow the log-normal distribution. When a model is built to satisfy all the above conditions it can be used to generate any number of sequences of the hydrologic variable of any desired length. For all practical purposes, there is no difference between the generated sequences and the observed sequence and they are as likely to occur as the observed sequence and they can be used wherever an observed sequence is used. The only difference then is that the generated sequence is not actually observed. Hence they are called *synthetic sequences*.

11.2. CLASSIFICATION OF MODELS

The various models used to generate synthetic sequences of any hydrologic variable may be broadly classified, based on three different criteria as given below :

(*i*) Based on whether we are dealing with single variable or more than one variable.

(*ii*) Based on the time interval used in descretising the series.

(*iii*) Based on the choice of the model for the stochastic component.

As per the first classification the models may be called *univariate models* or *multivariate models*. A univariate model is one which deals with only one variable. In a multivariate model we deal simultaneously with more than one variable. In hydrology, as we will be interested in modelling a hydrologic variable at any specific site or sites these are more popularly known as *single site models* and *multi-site models.*

As per the second classification a model may be called *annual model* or *seasonal model* depending on whether the sampling time interval of the series is a year or a season which is less than a year, say, a month, a week, or a day. As a combination of the first two classifications we may further have *single site annual models,* and *single site seasonal models* and *multi site annual models* and *muti site seasonal models.*

As per the third classification we may have autoregressive (*AR*) models, moving average (*MA*) models, autoregressive moving average (*ARMA*) models, autoregressive integrated moving average models (*ARIMA*) models, fractional Gaussian noise (fGn) models, etc.

Though the models that are described in this chapter are most commonly used to generate flow sequences it is to be emhasised here that they are equally useful to generate sequences of any other hydrologic variable, provided, of course, that the process under consideration is properly understood and a correct model is chosen for generation.

11.3. UNIVARIATE ANNUAL MODELS WITH NORMAL DISTRIBUTION

In the case of run off process, depending on the ground water storage and permeability characteristics of the basin, considerable amount of rainfall may become ground water and it takes a longer time for this ground water to appear as run off in the stream. Thus in the run off volume of any year there will be some contribution due to the rainfall of the previous year. This is especially true in the case of large basins. So, even if the rainfall of the current year and hence its contribution to the run off of the current year is small, it is likely that the run off of the currant year may still be large if the rainfall in the previous year and the corresponding run off of the previous year are large and *vice-versa*. Because of this storage effect of the basin, we find that in the case of annual run off series low flows tend to follow low flows and high flows tend to follow high flows. This storage effect, variously known as the *memory* of the process or the *persistence*, is truly reflected in the first order *AR* model as the value of the variable at any time in this model is dependent in part on the value of the variable in the previous time. This may also be noticed in the correlogram of the observed flows which shows a sort of exponential decay. Therefore it is often the $AR(1)$ model which is the choice to describe the annual flow series. If the memory extends for two time periods perhaps AR model (2) may have to be tried. But in hydrology it is only rarely that $AR(2)$ model may prove to be better choice than $AR(1)$ model for annual flows. Also the annual flow series is usually assumed to be stationary since there is no periodic or cyclic component. Hence AR models are often found to be suitable for describing the annual flow series. In the light of central limit theorem the annual flows are gerenally assumed to follow normal distribution. Hence the flows generated by the stochastic model should also follow normal distribution.

11.3.1. First order AR model

This model as defined by Eq. (9.27) is written as

$$(X_t - \mu_x) = \alpha_1 (X_{t-1} - \mu_x) + \xi_t \qquad ...(11.1)$$

where μ_x is the mean of the process X_t and ξ_t is an independent process with mean equal to zero. It has already been shows that α_1 is equal to the first serial correlation coefficient of the process ρ_1. From Eq. (9.26) the variance of the independent process ξ_t denoted by σ_ξ^2 and the variance of the process X_t denoted by σ_x^2 are related as

$$\sigma_\xi^2 = \sigma_x^2 (1 - \rho_1^2) \qquad ...(11.2)$$

Thus we see that the independent process ξ_t in the model should have mean equal to zero and the variance given by Eq. (11.2). If we consider another independent process Z_t with mean equal to 0 and variance equal to 1 and if we define

$$\xi_t = Z_t \cdot \sigma_x \sqrt{(1 - \rho_1^2)} \qquad ...(11.3)$$

then ξ_t will have the required variance specified by Eq. (11.2). Hence the first order AR model may now be written as

$$(X_t - \mu_x) = \rho_1 (X_{t-1} - \mu_x) + Z_t \sigma_x \sqrt{(1 - \rho_1^2)} \qquad ...(11.4)$$

Replacing the population parameters by the corresponding estimates obtained from the sample function, the $AR(1)$ model to generate the annual flows will take the form

$$x_t = \bar{x} + r_1 (x_{t-1} - \bar{x}) + z_t s_x \sqrt{(1 - r_1^2)} \qquad ...(11.5)$$

where $\bar{x}$ is the mean of the observed flows, s_x is the standard deviation of the observed flows, r_1 is the first autocurrelation coefficient, z_t is an independent random deviate with mean 0 and variance 1, and x_t is the flow generated in any year t.

The first term on the right hand side of the above equation is the mean of the annual flows. The second term represents the dependence of the flow in the year t on the flow in the previous year $(t - 1)$. The third term introduces the required variance into the generated sequence. If the annual flows are normally distributed the random variate z_t should also be normally distributed with mean zero and variance unity. Because in such a case the second and third terms of the generating scheme given by Eq. (9.6) are both random variates following normal distribution and the sum of two normal variables is also normally distributed.

The reader may be wondering at this stage as to how to produce the independent random deviate z_t which is N (0, 1). The techniques required to generate the independent series z_t not only of N (0, 1) but also of other distributions are described in detail in Chapter 13.

Example 11.1. *The annual run off of a river (in tmcft) for a period of 20 years are given in Example 6.2. Formulate an AR(1) model for this series and generate a synthetic flow sequence of 12 years. (The chain of 12 independent normal random deviates may be taken as 0.2619, – 0.7176, – 0.5749, – 0.0878, 0.7111, – 1.7767, 0.3506, 2.1505, – 0.9114, 0.2893, 0.1152 and 0.5870).*

Sol. The $AR(1)$ model is

$$x_t = \bar{x} + r_1 (x_{t-1} - \bar{x}) + z_t \,.\, s_x \,.\, \sqrt{(1 - r_1^2)}$$

The values of $\bar{x}$ and s_x are directly taken from the solution of Example 6.2. Thus $\bar{x}$ = 197.49 and s_x = 101.94. The first serial correlation coefficient r_1 is computed using Eq. (9.8) with the given annual flows. It is obtained as 0.4186. So, r_1 = 0.4186.

$$s_x \,.\, \sqrt{(1 - r_1^2)} = 101.94 \times \sqrt{(1 - 0.4186^2)} = 92.58$$

$$\therefore \quad x_t = 197.49 + 0.4186\,(x_{t-1} - 197.49) + 92.58\, z_t$$

The generated flow in the first year is obtained as

$$x_1 = 197.49 + 0.4186\,(x_0 - 197.49) + 92.58\, z_1.$$

where x_0 is called the seed of the generation. Usually the seed x_0 is taken to be $\bar{x}$ itself. The iterative procedure of the generation is clear from Table 11.1.

Table 11.1. Generation of Synthetic Annual Streamflows of Example 11.1

t	x_{t-1}	$(x_{t-1}-\bar{x})$	$r_1(x_{t-1}-\bar{x})$	z_t	$z_t s_x \sqrt{(1-r_1^2)}$	$x_t = \bar{x} + (4) + (6)$
(1)	(2)	(3)	(4)	(5)	(6)	(7)
1	197.49	0	0	0.2619	24.25	221.74
2	221.74	24.25	10.15	– 0.7196	– 66.62	141.02
3	141.02	– 56.47	– 23.54	– 0.5749	– 53.22	120.73
4	120.73	– 76.76	– 31.99	– 0.0878	– 8.13	157.36
5	157.36	– 40.13	– 16.80	0.7111	65.83	246.52
6	246.52	49.03	– 20.52	– 1.7767	– 164.49	53.52
7	53.52	– 143.97	– 60.27	0.3506	32.46	169.68
8	169.68	– 27.81	– 11.64	2.1505	199.09	384.94
9	384.94	187.45	78.47	– 0.9114	– 84.38	191.58
10	191.58	– 5.91	– 2.46	0.2893	26.78	221.81
11	221.81	24.32	10.14	0.1152	10.67	218.3
12	218.30	20.81	8.71	0.5870	54.34	260.54

The generation may be continued as long as it is required. When the flow is assumed to be normally distributed, it is possible that negative flows may be generated during some periods. This is physically unacceptable. In such cases it is recommended that the negative flow be used in the generation of subsequent flow and later all negative flows be made equal to zero or a small positive value. This difficulty does not arise when the generation is based on either lognormal or gamma distribution (to be discussed later).

To eliminate the influence of the seed of generation, the sequence is generated for a longer period than required and initial segment which is in excess is dropped from further usage. That means, if the required length of the generated sequence is, say 50 years, we generate the flows for about 60 or 65 years and drop the first 10 or 15 years of generated record. This is known as *warming up*. But when we are interested in more than one sequence, the warming up is done only for the first sequence. For the subsequent sequences the seed is taken as the last flow generated in the sequence previous to it.

11.3.2. Second Order AR Model

It is given by

$$(X_t - \mu_x) = \alpha_1 (X_{t-1} - \mu_x) + \alpha_2 (X_{t-2} - \mu_x) + \xi_t.$$

From Eq. (9.26), we have

$$\sigma_\xi^2 = \sigma_x^2 (1 - \alpha_1 \rho_1 - \alpha_2 \rho_2)$$

Similarly from Eqs. (9.34) and (9.35) the relations between the parameters of the process α_1 and α_2 and the first two autocorrelation coefficients ρ_1 and ρ_2 are

$$\alpha_1 = \frac{\rho_1 - \rho_1\rho_2}{1 - \rho_1^2} \quad \text{and} \quad \alpha_2 = \frac{\rho_2 - \rho_1^2}{1 - \rho_1^2}$$

On substitution of α_1 and α_2 in terms of ρ_1 and ρ_2 and simplification, we get

$$\sigma_\xi^2 = \sigma_x^2 \left[1 - \frac{(\rho_1^2 + \rho_2^2 - 2\rho_1^2\rho_2)}{(1 - \rho_1^2)} \right]$$

or $$\sigma_\xi^2 = \sigma_x^2 (1 - R^2)$$

where $$R^2 = (\rho_1^2 + \rho_2^2 - 2\rho_1^2\rho_2)/(1 - \rho_1^2) \quad ...(11.6)$$

and R is called the coefficient of multiple correlation.

Replacing the population parameters by their corresponding sample estimates, the generating scheme of the $AR(2)$ model is written in the form

$$x_t = \bar{x} + a_1 (x_{t-1} - \bar{x}) + a_2 (x_{t-2} - \bar{x}) + z_t \cdot s_x \cdot \sqrt{1 - R^2} \quad ...(11.7)$$

where $\bar{x}$ and s_x are the mean and standard deviation of the observed series and a_1, a_2 and R are given by

$$a_1 = \frac{(r_1 - r_1 r_2)}{(1 - r_1^2)} \quad ...(11.8)$$

$$a_2 = \frac{(r_2 - r_1^2)}{(1 - r_1^2)} \quad ...(11.9)$$

$$R^2 = \frac{(r_1^2 + r_2^2 - 2r_1^2 r_2)}{(1 - r_1^2)} \quad ...(11.10)$$

As usual z_t is an independent random variate following $N(0, 1)$. The $AR(2)$ model of Eq. (11.7) may be used in its present form only when the annual flow are normally distributed. If X_t is not normally distributed but follows some skewed distribution, the treatment of ξ_t becomes exceedingly complex.

Example 11.2. *Assuming that an AR(2) model provides a satisfactory fit to the annual flow of a river whose mean and standard deviation are estimated to be 525 Mm³ and 140 Mm³, and the first and second correlation coefficients to be 0.458 and – 0.004 generate annual flows for 3 years using the chain of normal random deviates of Example 11.1.*

Sol. We have $\bar{x} = 525$; $s_x = 140$; $r_1 = 0.458$; $r_2 = -0.004$

The generating scheme is given by

$$x_t = \bar{x} + a_1 (x_{t-1} - \bar{x}) + a_2 (x_{t-2} - \bar{x}) + z_t \,.\, s_x \sqrt{1 - R^2)}$$

From Eqs. (11.8), (11.9) and (11.10)

$$a_1 = \frac{r_1 - r_1 r_2}{1 - r_1^2} = \frac{0.458 - 0.458 \times (-0.004)}{1 - 0.458^2}$$

$$= \frac{0.458 + 0.001832}{0.790236} = 0.582$$

$$a_2 = \frac{r_2 - r_1^2}{1 - r_1^2} = \frac{-0.004 - 0.458^2}{0.790236} = -0.271$$

$$R^2 = \frac{r_1^2 + r_2^2 - 2r_1^2 r_2}{1 - r_1^2}$$

$$= \frac{0.458^2 + (-0.004)^2 - 2 \times 0.458^2 \times (-0.004)}{0.790236}$$

$$= 0.2676$$

$$\sqrt{(1 - R^2)} = \sqrt{(1 - 0.2676)} = 0.856$$

$$s_x \sqrt{(1 - R^2)} = 140 \times 0.856 = 119.84$$

$$x_t = 525 + 0.582 (x_{t-1} - 525)$$
$$- 0.271 (x_{t-2} - 525) + 119.84 z_t$$

In this scheme to generate x_1 two seeds x_0 and x_{-1} are needed. Both are assumed to be equal to $\bar{x}$.

$$\therefore \quad x_1 = 525 + 0 - 0 + 119.84 \times 0.2619 = 556.39$$

$$x_2 = 525 + 0.582 \times (556.39 - 525)$$
$$- 0 + 119.84 \times (-0.7196)$$
$$= 457.03$$

$$x_3 = 525 + 0.582 \times (457.03 - 525)$$
$$- 0.271 \times (556.39 - 525) + 119.84 \times (0.5749)$$
$$= 408.04$$

The other flows x_4, x_5, ... etc., can be generated for any required length in a similar manner.

11.3.3. ARMA Model

A general mixed autoregressive and moving average process is represented by Eq. (9.42). As mentioned earlier, a mixed process of considerable importance is *ARMA*(1, 1) process, which may be represented, after dropping the suffix for the parameters of the process α and β, by

$$(X_t - \mu_x) = \alpha (X_{t-1} - \mu_x) + \xi_t - \beta \xi_{t-1} \qquad \text{...(11.11)}$$

The ξ_t process should have the variance given by

$$\sigma_\xi^2 = \frac{\sigma_x^2 (1 - \alpha^2)}{(1 + \beta^2 - 2\alpha\beta)} \qquad \text{...(11.12)}$$

If Z_t is an independent process with mean 0 and variance 1, then $\xi_t = Z_t \times \sqrt{\sigma_x^2 (1 - \alpha^2)/(1 + \beta^2 - 2\alpha\beta)}$ will have the desired variance. Accordingly the model takes the form

$$(X_t - \mu_x) = \alpha (X_{t-1} - \mu_x) + (Z_t - \beta Z_{t-1}) \sqrt{\frac{\sigma_x^2 (1 - \alpha^2)}{(1 - \beta^2 - 2\alpha\beta)}} \qquad \text{...(11.13)}$$

One method of estimating α and β is to use Eq. (9.45) and Eq. (9.46) with the known values of ρ_1 and ρ_2. First α is obtained as ρ_2/ρ_1 from Eq. (9.46) and then substituting α and ρ_1 in Eq. (9.45) β is computed. But such an estimate can be taken as only an initial estimate which shall be improved upon using graphical procedure or non-linear estimation procedure which are not described here. The stationarity and invertibility conditions that $-1 < \alpha < 1$ and $-1 < \beta < 1$ will define the parameter space for the search.

Replacing the population parameters by their corresponding sample estimates, the *ARMA*(1, 1) model to generate the synthetic sequence may be written as

$$x_t = \bar{x} + a (x_{t-1} - \bar{x}) + T (z_t - b z_{t-1}) \qquad \text{...(11.14)}$$

where

$$a = r_2/r_1 \qquad \text{...(11.15)}$$

$$r_1 = \frac{(1 - ab)(a - b)}{(1 + b^2 - 2ab)} \qquad \text{...(11.16)}$$

$$T = s_x \times \sqrt{\frac{(1 - a^2)}{(1 + b^2 - 2ab)}} \qquad \text{...(11.17)}$$

Because of the typical characteristics of its correlogram, *ARMA*(1, 1) model may be a better choice when preservation of *Hurst coefficient* (to be discussed in later sections) is important.

Example 11.3. *The estimates of mean, standard deviation, first and second serial correlation coefficients of the annual run off of a stream are 1210 Mm^3, 570 Mm^3, 0.535 and 0.463 respectively. Assuming that an ARMA (1, 1) model is a good fit to describe the flows, obtain the initial estimate of the parameters and hence generate a sequence of 3 annual flows using the following independent standard normal deviates + 1.123, – 0.821, – 0.342.*

Sol. Given $\bar{x} = 1210$; $s_x = 570$; $r_1 = 0.535$; $r_2 = 0.463$

$$a = \frac{r_2}{r_1} = \frac{0.463}{0.535} = 0.8654$$

$$r_1 = \frac{(1 - ab)(a - b)}{(1 + b^2 - 2ab)}$$

$$0.535 = \frac{(1 - 0.8654b)(0.8654 - b)}{(1 + b^2 - 1.7308b)}$$

$$0.3304b^2 - 0.8229\,b + 0.3304 = 0$$

$$b = 1.987 \quad \text{or} \quad 0.503$$

Since 1.987 is not acceptable, $b = 0.503$

$$T = s_x \sqrt{\frac{1 - a^2}{1 + b^2 - 2ab}}$$

$$= 570 \sqrt{\frac{1 - 0.8654^2}{(1 + 0.503^2 - 2 \times 0.8654 \times 0.503)}}$$

$$= 461.86.$$

$$\therefore \; x_t = 1210 + 0.8654\,(x_{t-1} - 1210) + 461.86\,(z_t - 0.503\,z_{t-1})$$

The first flow is generated by letting $x_0 = \bar{x}$ and $z_0 = \bar{z}$, that is zero.

$$x_1 = 1210 + 0.8654 \times 0 + 461.86\,(1.123 - 0.503 \times 0)$$
$$= 1728.67$$

$$x_2 = 1210 + 0.8654\,(1728.67 - 1210) + 461.86\,(-0.821 - 0.503 \times 1.123)$$
$$= 1018.78$$

$$x_3 = 1210 + 0.8654\,(1018.78 - 1210) + 461.86\,(-0.342 + 0.503 \times 0.821)$$
$$= 1077.3$$

The other flows x_4, x_5, ... for any desired length can be computed following the same procedure.

11.4. UNIVARIATE ANNUAL MODELS WITH NON-NORMAL DISTRIBUTION

The models discussed in the previous section are assumed to be applied to discrete series which are second-order stationary or which are transformed to second order stationarity. It is implied that the distribution of X_t and of ξ_t is normal. This approximation is generally sufficient in the case of annual series. However, modifications may be necessary when dealing with hydrological series observed on monthly basis and other more closely spaced data. Sometimes even annual series may also be skewed.

If X_t is non-normally distributed with skewness γ_x, then tne skewness γ_ξ to be incorporated in the random component ξ_t depends on the model and its parameters. Consider for example $AR(1)$ model.

$$(X_t - \mu_x) = \alpha\,(X_{t-1} - \mu_x) + \xi_t$$

Cubing the above equation on both sides and taking expectations

$$E[(X_t-\mu_x)^3] = \alpha^3 E[(X_{t-1}-\mu_x)^3] + 3\,\alpha_1^2 E[(X_{t-1}-\mu_x)^2 \xi_t] + 3\,\alpha_1 E[(X_{t-1}-\mu_x)\,\xi_t^2] + E\left[\xi_t^3\right]$$

Now $E[(X_t-\mu_x)^3] = \mu_{3x}$; $E[(X_{t-1}-\mu_x)^3] = \mu_{3x}$

$E[(X_{t-1}-\mu_x)^2 \xi_t] = 0$; $E[(X_{t-1}-\mu_x)\,\xi_t^2] = 0$

Because X_{t-1} and ξ_t are independent.

$$E[\xi_t^3] = \mu_{3\xi} \quad \text{since} \quad \mu_\xi = 0.$$

$$\therefore \qquad \mu_{3x} = \alpha^3 \mu_{3x} + \mu_{3\xi}\,.$$

Which gives $\mu_{3\xi} = (1-\alpha_1^3)\,\mu_{3x}$

By definition, $\gamma_\xi = \dfrac{\mu_{3\xi}}{(\sigma_\xi^2)^{3/2}} = \dfrac{(1-\alpha_1^3)\,\mu_{3x}}{(\sigma_\xi)^3}$

But in $AR(1)$ model, $\alpha_1 = \rho_1$ and $\sigma_\xi = \sigma_x \sqrt{(1-\rho_1^2)}$

Thus
$$\gamma_\xi = \frac{(1-\rho_1^3)\,\mu_{3x}}{(1-\rho_1^2)^{3/2}\,\sigma_x^3}$$

Again by definition $\dfrac{\mu_{3x}}{\sigma_x^3} = \gamma_x$, the skewness of X_t.

$$\gamma_\xi = \frac{\gamma_x (1-\rho_1^3)}{(1-\rho_1^2)^{3/2}} \qquad \text{...(11.18)}$$

For estimation purpose γ_x and ρ_1 in Eq. (11.18) may be replaced by g_x and r_1.

Gamma distribution. When the annual series follow a gamma distribution with skewness coefficient g_x, the skewness coefficient of the random component g_ξ may be computed from the following equation.

$$g_\xi = \frac{g_x (1-r_1^3)}{(1-r_1^2)^{1.5}} \qquad \text{...(11.19)}$$

The $AR(1)$ generation scheme is then given by

$$x_t = \bar{x} + r_1 (x_{t-1} - \bar{x}) + p_t \,.\, s_x \,.\, \sqrt{(1-r_1^2)} \qquad \text{...(11.20)}$$

where p_t is obtained from

$$p_t = \frac{2}{g_\xi}\left(1 + \frac{g_\xi z_t}{6} - \frac{g_\xi^2}{36}\right)^3 - \frac{2}{g_\xi} \qquad \text{...(11.21)}$$

As in the earlier model z_t is an independent normal random deviate with mean 0 and variance 1.

Lognormal distribution. When the series are observed to follow lognormal distribution, first it is subjected to logarithmic transformation given by

$$y = \ln (x - a) \quad \text{...(11.22)}$$

where the parameter, a representing the lower bound could be negative or positive. Then the statistical parameters of the transformed series, namely, mean, standard deviation and first serial correlation coefficient denoted by $\bar{y}$, s_y and r_{1y} are estimated. Now the y series is generated using the $AR(1)$ model

$$y_t = \bar{y} + y_{1y} (y_{t-1} - \bar{y}) + z_t \,.\, s_y \,.\, \sqrt{(1 - r_{1y}^2)} \quad \text{...(11.23)}$$

where z_t is an independent normal deviate with mean and variance equal to 0 and 1 respectively. Now the synthetic sequence of x-series is obtained through detransformation

$$x_t = a + e^{y_t} \quad \text{...(11.24)}$$

The generating scheme given by Eq. (11.23) preserves the statistical structure only in the transformed logarithmic domain and not necessarily in the original series and this discrepancy could be significant. Matalas has suggested that the following relations which exist between the original and the transformed series can be used to estimate the parameters of the transformed series.

$$\bar{x} = a + \exp (s_y^2/2 + \bar{y}) \quad \text{...(11.25)}$$

$$s_x^2 = \exp (2\bar{y} + s_y^2) \,.\, \exp (s_y^2 - 1) \quad \text{...(11.26)}$$

$$g_x = \{\exp (3s_y^2) - 3 \exp (s_y^2) + 2\}/\{\exp (s_y^2 - 1)\}^{3/2} \quad \text{...(11.27)}$$

By substituting the known values of g_x, s_x and $\bar{x}$ in the left hand side of Eqs. (11.27), (11.26) and (11.25) respectively, the values of s_y^2, $\bar{y}$ and a are obtained in the same order. Now by substituting r_{1x} in Eq. (11.28), r_{1y} can be evaluated.

$$r_{1x} = \exp (s_y^2 \,.\, r_{1y} - 1)/\{\exp (s_y^2) - 1\} \quad \text{...(11.28)}$$

Then the parameters $\bar{y}$, s_y and r_{1y} are used in Eq. (11.23) to generate the y series which are detransformed to x series using Eq. (11.24), along with the estimated value of a. Alternatively, the x series can be generated directly using the following scheme.

$$x_t = a + e^{\bar{y}(1 - r_{1y})} \,.\, (x_{t-1} - a)^{r_{1y}} \,.\, e^{z_t \,.\, s_y \,.\, \sqrt{(1 - r_{1y}^2)}} . \quad \text{...(11.29)}$$

11.5. UNIVARIATE ANNUAL MODELS OBEYING HURST'S LAW

Let $x_1, x_2, \ldots\ldots, x_n$ be a sequence of inflows into a reservoir over a period of n year. Let $\bar{x}$ be the mean of these flows. If the rate of withdrawls from the reservoir is equal to $\bar{x}$ every year, then in the first k years the cumulative inflow into the reservoirs would be $\sum_{t=1}^{k} x_t$ and the cumulative withdrawl would be $k\bar{x}$. The difference between these two quantities, denoted by D_k, is known as the *cumulative departure* and represents the *excess* or the *deficiency* relative to the total amount of water withdrawn from the reservoir upto kth year. That is

$$D_k = \sum_{t=1}^{k} x_t - k\bar{x} \qquad \ldots(11.30)$$

Fig. 11.1 shows schematically the cumulative departures.

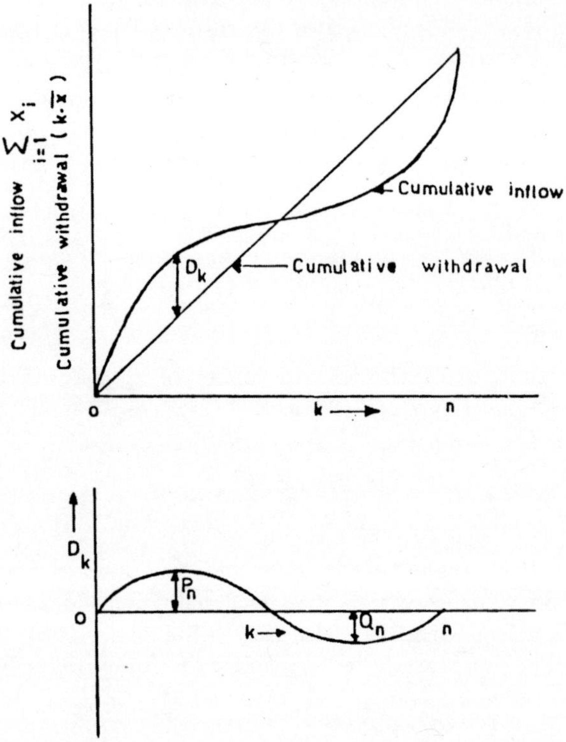

Fig. 11.1. Schematic representation of cumulative departure.

Let P_n denote the maximum of all D_k for $0 \le k \le n$, and Q_n denote the minimum of all D_k for $0 \le k \le n$ which represent the largest excess and greatest deficiency during the n years respectively. The difference of these two quantities denoted by R_n is known as the *range of cumulative departures* and is denoted by R_n. That is

$$R_n = P_n - Q_n \qquad ...(11.31)$$

The range R_n represents the storage capacity of an ideal reservoir to maintain a uniform withdrawl equal to $\bar{x}$. Let s_n be the standard deviation of the inflows. Then the quantity R_n/s_n is called the *rescaled range*.

Hurst, in his extensive analysis, computed, R_n for some 800 time series of annual values of streamflows, precipitation, temperature, tree rings, mud varves, etc., with the length of the series, n, varying from 40 to 2000 years. He then found the rescaled range to vary with n as

$$\frac{R_n}{s_n} \propto n^h \qquad ...(11.32)$$

with value of h, known as the *Hurst's coefficient,* lying in the range 0 to 1. To determine h, Hurst defined

$$\frac{R_n}{s_n} = \left(\frac{n}{2}\right)^k \qquad ...(11.33)$$

and computed k, which represents an estimate of h, for each of the time series. Over all phenomena k was found to have an average value of 0.73 and a standard deviation of 0.08. Eq. (11.33) is known as *Hurst's law*. For a purely random process with n large, it has been theoretically established that $h = 1/2$. The departure of the exponent k in Eq. (11.33) from 0.5 is called the *Hurst phenomenon*. In other words, Hurst's findings suggest that, for a given n, the size of the reservoir has to be larger than the capacity corresponding to a purely random series of inflow. The autoregressive models discussed hitherto are also characterised by $h = 0.5$. Therefore, the use of synthetic flow sequences based on *AR* models, if used in reservoir, design will underestimate the storage requirement resulting in economic loss. Some of the possible explanations of the Hurst phenomenon include non-stationarity and long memory.

Mandelbrot and Wallis proposed a stationary stochastic model called *fractional Gaussian noise (fGn) model* which obeys the Hurst's law with h lying in the range 0 to 1 thus accounting for the Hurst phenomenon. The form of autocorrelation function of fGn implies the existence of infinite memory which in turn gives rise to the phenomenon of long-term persistence. The theoretical autocorrelation function of fGn process is given by

$$\rho_k = [\ |\ k+1\ |^{2h} - 2\ |\ k\ |^{2h} + |\ k-1\ |^{2h}]/2 \qquad ...(11.34)$$

The spectral density function $\gamma(f)$ of *fGn* is of the form $\gamma(f) \sim \rho^{1-2h}$ where $0 < h < 1$ with the condition that $h \neq 0.5$. The term *fractional* arises from the type of this function since h is always fractional. For white noise, h takes on the (asymptotic) value of 0.5 which means that, as mentioned in Sec. 10.5, its spectrum is horizontal.

Some investigators have shown that ARIMA models can be regarded as approximations to *fGn*. Specifically ARIMA (1, 0, 1) which is equivalent to ARMA (1, 1) has been shown to be capable of describing Hurst phenomenon in some cases. The ARMA (1, 1) model can exhibit distinctly different long-term properties compared to AR models as seen from the comparison shown in Fig. 11.2.

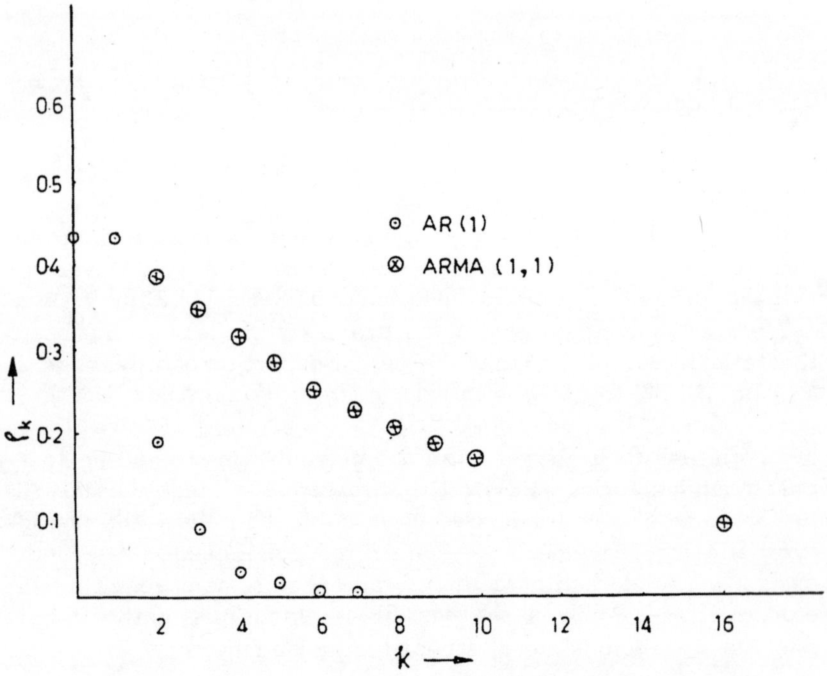

Fig. 11.2. Comparison of ρ_k of ARMA(1, 1) process with $\alpha = 0.9$ and $\beta = 0.64$ and AR(1) process with $\alpha = 0.428$.

An ARMA (1, 1) process with $\alpha = 0.9$ and $\beta = 0.64$ will have ρ_1 [as computed from Eq. (9.45)] equal to 0.428. Using Eq. (9.46) the values of ρ_k, $k = 1, 2, ...$ are evaluated and its correlogram is constructed. The correlogram of AR (1) process with $\rho_1 = 0.428$ is also evaluated using the relation that $\rho_k = \rho_1{}^k$, for $k = 2, 3, \ldots\ldots$ and is plotted on the same diagram as shown in Fig. 11.2. Though ρ_1 is same for both the processes, it can be observed that the correlogram of ARMA (1, 1) process exhibits long term memory characteristics which are distinctly different from that of AR (1) process.

The ARMA (1, 1) model has already been discussed in detail in Section 11.3.3.

11.5.1. Fractional Gaussian noise models

These models are characterised by infinite memory. For practical adoptibility they have been approximated by a finite but large memory. Mandelbrot and Wallis proposed approximations to *fGn* which they referred to as Type I and Type II. The type I approximation proved expensive computationally. Now we shall present only type II approximation.

Type II approximation to fGn. The series based on type II approximation to *fGn* can be generated by

$$x_t = (h - 0.5) \sum_{u = t - M}^{t-1} (t - u)^{h-1.5} \cdot z_u \qquad \ldots(11.35)$$

where h is the Hurst's coefficient, M is the memory of the process and z_u for $u = 0, \pm 1, \pm 2, \pm 3, \ldots\ldots$ are independent standard normal variates. Hence for $t = 1$, we have

$$x_1 = (h - 0.5)\,(1^{h - 1.5} \cdot z_0 + 2^{h - 1.5} \cdot z_{-1} + 3^{h - 1.5} \cdot z_{-2} + \ldots\ldots + M^{h - 1.5} \cdot z_{-M+1})$$

and for $t = 2$

$$x_2 = (h - 0.5)\,(1^{h - 1.5} \cdot z_1 + 2^{h - 1.5} \cdot z_0 + 3^{h - 1.5} \cdot z_{-1} + \ldots\ldots + M^{h - 1.5} \cdot z_{-M+2})$$

In this way the series x_t, $t = 1, 2, \ldots\ldots, n$ can be generated which preserves the Hurst coefficient. That means the rescaled ranges of x_t series will be proportional to n^h. It may be noted that, in order to generate a single *fGn* variate, we require M independent normal variates z_u. The recommended value of M is 20,000. In order to generate the next *fGn* variate, the last independent normal variate is discarded and a new independent normal variate is produced but the same set of M weights is used.

Filtered fGn model. The type II approximation of *fGn* will be lacking in high frequencies. Because such effects are undesirable, Matalas and Wallis introduced a filtered *fGn* model. For this purpose Eq. (11.35) is modified as follows.

$$x_t = (h - 0.5) \sum_{u = pt - M}^{pt - 1} (pt - u)^{h-1.5} \cdot z_u \qquad \text{...(11.36)}$$

where p is a filtering parameter which is an arbitrarily chosen integer greater than 1. In applying Eq. (11.36), the same set of weights is used as for Eq. (11.35). The difference is that to generate x_{t-1}, the last p independent normal random variates used in generating x_t are discarded and p new variates are added to the start of the sequence of z_u.

Fast fractional Gaussian noise model. The main draw back of the type I, or type II, or filtered type II approximations is the excessive computer time required. To remedy the situation Mandelbrot postulated a fast fractional Gaussian noise (*ffGn*) model. In this, the model has two additive components. That is,

$$x_t = x_t^{\,L} + x_t^{\,H} \qquad \text{...(11.37)}$$

where $x_t^{\,L}$ is to account for the *low* frequency effects and $x_t^{\,H}$ is to account for the *high* frequency effects. Now $x_t^{\,L}$ is given by

$$x_t^{\,L} = \sum_{i=1}^{p} W_i X_t\,(GM/i, B) \qquad \text{...(11.38)}$$

where $X_t\,(GM\,/\,i, B)$ is a standard $AR(1)$ normal process with ρ_1 equal to $(-B)^{-i}$ and with standard normal independent variates as random component. The parameter B takes a value in the suggested range of 2 to 4, but computer experiments show that the variations within this range are not over-sensitive to a design parameter such as reservoir capacity. The weights W_i are given by

$$W_i^{\,2} = \frac{h\,(2h-1)\,(B^{1-h} - B^{h-1})B^{-2(1-h)i}}{\Gamma(3-2h)} \qquad \text{...(11.39)}$$

It is originally suggested that parameter p should be functional to the desired sample size n, *i.e.* $p = f(n)$, but p = 15 to 20 is found to be sufficient in practice.

The component $x_t^{\,H}$ is a separate $AR(1)$ normal process with zero mean. Its variance is $1 - B^{h-1}\,(2h^2 - h)/\Gamma(3-2h)$ and its lag one autocorrelation coefficient is given by $2^{2h-1} - 1 + \sum_{i=1}^{p} W_i\,(1 - r_i) - B^{-(1-h)}\,h\,(2h-1)/\Gamma(3-2h)$. The reduction in computer time is such that 2000 variates of *ffGn* can be generated in the time taken to generate one variate of type II *fGn* with M = 20000.

11.6. UNIVARIATE SEASONAL MODELS

As we have been repeatedly pointing out, any hydrologic series observed with a sampling time interval of less than a year would inevitably be non-stationary in structure because of the cyclic component with a period of one year introduced into it as a result of the astronomic cycle. Therefore, seasonal models have to take care of this non-stationarity in one way or the other. There are basically two approaches which have been attempted successfully. In the first approach the model is basically structured as a non-stationary model (or shall we call it cyclic stationarity ?) which tries to preserve the various statistical parameters like mean, standard deviation, the correlation co-efficient etc. which vary from season to season. In the second approach an appropriate complex periodic deterministic function is fitted to account for the cyclicity in the variable to which a stationary stochastic component properly fitted by either *AR* or *MA* models is added to obtain synthetic seasonal flows. Either of the approaches has its own advantages and disadvantages and must be judiciously selected to suit the individual problem. The first approach is due to Thomas and Fiering and the latter is due to Yevjewich and Quimpo.

11.6.1. Thomas-Fiering Model

This model implicitly allows for the non-stationarity of seasonal data. It is written in the form

$$X_t = \mu_j + \frac{\rho_j \sigma_j}{\sigma_{j-1}} [X_{t-1} - \mu_{j-1}] + z_t \sigma_j \sqrt{1 - \rho_j^{\,2}} \quad ...(11.40)$$

where X_t and X_{t-1} are the flows during the time periods t and $(t-1)$ respectively, μ_j and σ_j are the mean and standard deviation of the flows in jth season respectively within a repetitive cycle of m seasons, ρ_j is the correlation co-efficient between the flows of jth and $(j-1)$th seasons, and z_t is a $N(0, 1)$. At any time step t, j is obtained as $j = t \bmod (m)$, which has the meaning that t is divided by m and the remainder is taken as the value for j. If $t \bmod (m) = 0$, then $j = m$. And when $j = 1, j - 1 = m$. Equation (11.40) may be written in a more lucid form as

$$X(i, j) = \mu_i + \frac{\rho_j \sigma_j}{\sigma_{j-1}} [X(i, j-1) - \mu_{j-1}] + z_t \cdot \sigma_j \sqrt{(1 - \rho_j^{\,2})} \quad ...(11.41)$$

where $i = 1, 2, \ldots\ldots$

$j = 1, 2, \ldots\ldots, m$

$t = (i - 1)\, m + j$

when $j = 1, X(i, j - 1) = X(i - 1, m)$

Here $X(i, j)$ is the flow in the jth season of ith year with j changing cyclically from 1 to m. If a season is a month, $m = 12$, for a week $m = 52$ and for a day $m = 365$ and so on. For any given season j, if we take the following expectations on both sides of Eq. (11.41), we can observe that

$$E[X(i, j)] = \mu_j$$

$$E[\{X(i, j) - \mu_j\}^2] = \sigma_j^2$$

$$E\left[\frac{\{X(i,j) - \mu_j\}}{\sigma_j} \frac{\{X(i,j-1) - \mu_{j-1}\}}{\sigma_{j-1}}\right] = \rho_j$$

These results ensure that Thomas-Fiering model preserves the non-stationary character of seasonal flows in mean, variance, and covariance of lag one. In practice the seasonal model of Eq. (11.41) is used by replacing the population parameters by their respective sample estimates obtained from the historical record. That is

$$x(i, j) = \bar{x}_j + b_j [x(i, j-1) - \bar{x}_{j-1}] + z_t s_j \sqrt{(1 - r_j^2)} \quad ...(11.42)$$

where $\bar{x}_j$ is the mean flow in season j.

s_j is the standard deviation of flows in season j

r_j is the correlation co-efficient between the flows of jth and $(j-1)$th season.

b_j is the regression co-efficient of flows in jth season with flows of $(j-1)$th season

$x(i, j)$ is the flow generated in the jth season of the ith year, with $j = 1, 2, ..., m$

$i = 1, 2, ...$

$t = (i-1)m + j$ and when $j = 1$; $x(i, j-1) = x(i-1, m)$

Let $Q(i, j)$ be the observed flow in jth season of ith year with an available record of n years. Then the above sample estimates are computed as given below for any season j.

$$\bar{x}_j = \frac{1}{n} \sum_{i=1}^{n} Q(i, j) \quad ...(11.43)$$

$$s_j = \sqrt{\frac{1}{(n-1)} \sum_{i=1}^{n} [Q(i,j) - \bar{x}_j]^2} \quad ...(11.44)$$

$$r_j = \frac{\frac{1}{n} \sum_{i=1}^{n} \left[\{Q(i,j) - \bar{x}_j\}\{Q(i,j-1) - \bar{x}_{j-1}\}\right]}{s_j \, s_{j-1}} \quad ...(11.45)$$

for $j = 2, 3, 4, ..., m$.

$$b_j = \frac{r_j s_j}{s_{j-1}} \quad \text{for } j = 2, 3, 4, \ldots\ldots, m \qquad \ldots(11.46)$$

and specifically (why ?)

$$r_1 = \frac{\dfrac{1}{n-1} \sum_{i=2}^{n} \left[\left\{ Q(i, 1) - \bar{x}_1 \right\} \left\{ Q(i-1, m) - \bar{x}_m \right\} \right]}{s_1 \cdot s_m} \qquad \ldots(11.47)$$

$$b_1 = \frac{r_1 \cdot s_1}{s_m} \qquad \ldots(11.48)$$

In order that the parameters in the model be estimated with reasonable accuracy, it is suggested that the Thomas-Fiering model be used only with caution if fewer than 12 years of data is employed.

When the Thomas-Fiering model is used to generate the seasonal flows with normal distribution, it is quite possible that, as in the case of annual flows with normal distribution, some times the generated flow may be negative. When such negative flows are generated the recommended practice is again to use these negative values in the generation of subsequent flows and replace all such negative flows by zero or a very small value once the generation is completed. When gamma or lognormal distribution is assumed for the seasonal flows this difficulty will not be encountered.

In the Thomas-Fiering model we notice that the flow in any season j is a sum of three terms. First term is the mean flow in that season. Second term is the regressed component on the flow in the previous season. The third term is a random component to reflect the desired variance. This model accounts for the persistence only upto one lag. Hence the Thomas-Fiering model may be viewed as a non-stationary first order AR model. This model can be applied directly in the form given in Eq. (11.42) drawing the random variate z_t from N (0, 1) provided the flows in all the seasons are normally distributed. When the flows follow a skewed distribution, which is usually the case with seasonal flows, certain modifications are required on the random component in order to preserve the skewness and the corresponding distribution. Suitable modifications for gamma and log-normal distributions are now discussed.

Gamma distribution. Let the seasonal flows be gamma distributed with the sample skewness coefficient of the flow in the jth season being denoted by g_j which can be evaluated from

$$g_j = \frac{\frac{1}{n}\sum_{i=1}^{n} [Q(i,j) - \bar{x}_j]^3}{s_j^{\,3}} \quad \text{...(11.49)}$$

Then the modified skewness coefficient for the season j, denoted by $g_j^{\,1}$, is computed as

$$g_j^{\,1} = \frac{g_j - r_j^{\,3}\, g_{j-1}}{\left[1 - r_j^{\,2}\right]^{1.5}} \quad \text{...(11.50)}$$

Now Equation (11.42) is employed to generate synthetic seasonal flows by replacing the independent normal random deviate z_t by p_t where

$$p_t = \frac{2}{g_j^{\,1}}\left[1 + \frac{g_j^{\,1} z_t}{6} - \frac{g_j^{\,1}}{36}\right]^3 - \frac{2}{g_j^{\,1}} \quad \text{...(11.51)}$$

This preserves the desired skewness coefficient in the seasonal flows which follow gamma distribution.

Lognormal distribution. When the seasonal flows are lognormally distributed, then their logarithms are normally distributed. Therefore, in this case, first the logarithm of the flows are generated using the random deviate from N (0, 1) and if we take the antilogarithm, we get the flows. If y is used to denote the logarithm of the seasonal flows the generation scheme would then be

$$y\,(i, j) = \bar{y}_j + b_{yj}\,[y(i, j-1) - \bar{y}_{j-1}] + z_t\, s_{yj}\,[(1 - r_{yj}^{\,2})]^{1/2} \quad \text{...(11.52)}$$

As already mentioned in the case of annual flows, if the parameters of the y series are estimated from the logarithms of the flows the resemblance of the stochastic structure will be only in the transformed log domain and not in the untransformed original flows. Following on the same lines as Matalas, who has suggested a mathematically correct approach in the case of annual flows which has been described earlier, Burges developed a procedure to estimate the parameters of the y series from the parameters of the original series as given below.

$$\bar{x} = e^{(\bar{y}_j + s_{yi}^{\,2}/2)} \quad \text{...(11.53)}$$

$$s_j^{\,2} = e^{(2s_{yj}^{\,2} + 2\bar{y}_j)} - e^{(s_{yj}^{\,2} + 2\bar{y}_j)} \quad \text{...(11.54)}$$

$$r_j = \frac{e^{(s_{yj-1}\cdot s_{yj}\, r_{yj})} - 1}{[e^{s^2_{yj-1}} - 1]^{1/2}\,.\,[e^{s^2_{yj}} - 1]^{1/2}} \quad \text{...(11.55)}$$

$$b_{yj} = \frac{r_{yj}\, s_{yj}}{s_{yj-1}} \quad ...(11.56)$$

The solution of the above three equations for each season j would provide the estimates of the parameters of the y series which, when used in Eq. (11.52), would ensure the statistical resemblance in the generated flows. Now the flow sequence itself is obtained from

$$x\,(i, j) = e^{[y(i, j)]} \quad ...(11.57)$$

Example 11.4. *Generate synthetic monthly flows for a period of 12 months using the statistical parameters given in Example 9.7 and Thomas Fiering model. Assume the flows in all the months are normally distributed and use the same chain of normal deviates given in example 11.1.*

Sol. To generate the flow in the first month of the first year we require the flow in the season previous to it which is $x\,(0, m)$ and this is the seed of generation taken to be $\bar{x}_m$ itself and in the present case $\bar{x}_{12} = 0.49$. The iterative steps involved in the generation are incorporated in Table (11.2).

11.6.2. Daily flow model

The number of parameters required in the Thomas-Fiering model are 3 m or 4 m depending on whether the skewness co-efficient is also to be preserved or not. Therefore, this model is quite convenient when the season is defined to be a month or half-a-year. The parameters required would be 36 or 48 in the former case and 6 or 8 in the latter case. On the other hand if the season is taken to be a week or a day, the number of parameters required is too large which is 156 or 208 in the case of weekly flows, and 1095 or 1460 in the case of daily flows. In such circumstances it becomes quite unwieldy to use the Thomas-Fiering model. It may, perhaps, be more convenient to describe the non-stationarity by complex periodic deterministic function as demonstrated by Quimpo and Yevjewich, which is now described.

When a correlogram is constructed for observed daily flows, it will inevitably show cyclicity with a period of 365 days or a fundamental frequency of one cycle/year. If the spectral density function is also estimated, it will further show that some more harmonics are significant. These significant harmonics may be required to express the periodic variability in mean and variance. In many a case, less than six harmonics are sufficient to account for about more than 90% of the explained variance. Since the fundamental period and fundamental frequency are clearly known we can straight away construct the periodogram or the line spectrum

Table 11.2. Generation of Monthly flows of Example 11.4 using Thomas-Fiering Model

i	j	$\bar{x}_j$	s_j	r_j	b_j	$x(i, j-1)$	$[x(i, j-1) - \bar{x}_{j-1}]b_j$	z_t	$z_t.s_j.\sqrt{1-r_j^2}$	$x(i, j) = (3) + (8) + (10)$
(1)	(2)	(3)	(4)	(5)	(6)	(7)	(8)	(9)	(10)	(11)
1	1	8.81	8.57	0.1563	5.15	0.49	0	0.2619	2.22	11.03
1	2	29.03	27.78	−0.0378	−0.12	11.03	−0.26	−0.7196	−19.98	8.79
1	3	54.94	59.36	0.1778	0.38	8.79	−7.69	−0.5749	−33.58	13.67
1	4	55.79	42.32	0.1802	0.13	13.67	−5.37	−0.0878	−3.65	46.77
1	5	32.52	21.91	−0.0388	−0.02	46.77	0.18	0.7111	15.57	48.27
1	6	7.04	9.14	0.1005	0.04	48.27	0.63	−1.7767	−16.16	−8.49*
1	7	2.98	1.79	0.7869	0.15	−8.49	−2.33	0.3506	0.39	1.04
1	8	2.08	0.80	0.8485	0.38	1.04	−0.74	2.1505	0.91	2.25
1	9	1.40	0.73	0.3395	0.31	2.25	0.05	−0.9114	−0.63	0.82
1	10	1.37	0.17	0.8201	1.31	0.82	−0.76	0.2893	0.19	0.80
1	11	1.07	0.95	0.0591	0.05	0.80	−0.03	0.1152	0.11	1.15
1	12	0.49	0.26	0.5164	0.14	1.15	0.01	0.5870	0.13	0.63

*Since this flow is negative it must be made equal to zero.

with $f_1 = 1$ cycle/year. Once the variation in mean and variance of daily flows is described by harmonic fitting we can obtain the residual series which are free from cyclicity. Further analysis is then carried out to identify the appropriate stationary model to fit the residual series. The various steps involved in the analysis and generation of daily flow series are discussed below in detail.

Let $x\ (i, j)$ be the observed flow in the jth day of the ith year, where $j = 1, 2, \ldots\ldots, 365$ and $i = 1, 2, \ldots\ldots, n$ and n being the number of years of record available. The mean and standard deviation of flows in jth day are computed from.

$$\bar{x}_j = \frac{1}{n} \sum_{i=1}^{n} x\ (i, j) \qquad \ldots(11.58)$$

$$s_j = \sqrt{\frac{1}{(n-1)} \sum_{i=1}^{n} [x(i, j) - \bar{x}_j]^2} \qquad \ldots(11.59)$$

The mean of the daily means $\bar{X}$, and the variance of the daily mean VX, and the mean of the daily standard deviations $\bar{S}$, and the variance of the daily standard deviations VS are then estimated from

$$\bar{X} = \frac{1}{365} \sum_{j=1}^{365} \bar{x}_j \qquad \ldots(11.60)$$

$$VX = \frac{1}{365} \sum_{j=1}^{365} [\bar{x}_j - \bar{X}]^2 \qquad \ldots(11.61)$$

$$\bar{S} = \frac{1}{365} \sum_{j=1}^{365} s_j \qquad \ldots(11.62)$$

$$VS = \frac{1}{365} \sum_{j=1}^{365} [s_j - \bar{S}]^2 \qquad \ldots(11.63)$$

Though the number of harmonics required for exact fitting is equal to $(365 - 1)/2 = 182$, it has been found from experience that the first 6 harmonics would give more than a reasonable fit in many cases. Therefore, the first six harmonic co-efficeints to fit the mean, namely a_k and b_k, and to fit the standard deviation namely c_k and d_k are computed for $k = 1, 2, \ldots\ldots, 6$ from

$$a_k = \frac{2}{365} \sum_{j=1}^{365} \bar{x}_j \cos\left(\frac{2\pi jk}{365}\right) \qquad \ldots(11.64)$$

$$b_k = \frac{2}{365} \sum_{j=1}^{365} \bar{x}_j \sin\left(\frac{2\pi jk}{365}\right) \qquad \text{...(11.65)}$$

$$c_k = \frac{2}{365} \sum_{j=1}^{365} \bar{s}_j \cos\left(\frac{2\pi jk}{365}\right) \qquad \text{...(11.66)}$$

$$d_k = \frac{2}{365} \sum_{j=1}^{365} s_j \sin\left(\frac{2\pi jk}{365}\right) \qquad \text{...(11.67)}$$

Now a decision may be taken on the number of harmonics to be used in the fitting. Let qm be the number of harmonics required to fit the periodicity in mean, and qs be the number of harmonics required to fit the periodicity in standard deviation. Then we select qm and qs such that

$$\sum_{k=1}^{qm} \frac{(a_k^2 + b_k^2)}{2} \geq 0.9\, VX \qquad \text{...(11.68)}$$

and

$$\sum_{k=1}^{qs} \left(\frac{c_k^2 + d_k^2}{2}\right) \geq 0.9\, VS \qquad \text{...(11.69)}$$

Generally qm is slightly more than qs. Now the mean and the standard deviation of the daily flows in any day j are represented by harmonic fitting as

$$\bar{x}_j' = \bar{X} + \sum_{k=1}^{qm} \left[a_k \cos\left(\frac{2\pi jk}{365}\right) + b_k \sin\left(\frac{2\pi jk}{365}\right)\right] \qquad \text{...(11.70)}$$

$$\bar{s}_j' = \bar{S} + \sum_{k=1}^{qs} \left[c_k \cos\left(\frac{2\pi jk}{365}\right) + d_k \sin\left(\frac{2\pi jk}{365}\right)\right] \qquad \text{...(11.71)}$$

After the harmonic analysis is over the residual series are obtained from

$$y_t = \frac{x(i,j) - \bar{x}_j'}{s_j'} \qquad \text{...(11.72)}$$

where $t = (i-1)365 + j$. The y series obtained from Eq. 11.72 will not have mean and standard deviation exactly equal to zero and unity respectively since the fitted values $\bar{x}_j'$ add s_j' are not exactly equal to their observed counterparts $\bar{x}_j$ and s_j. Nevertheless since the y series is free from cyclicity, it is assumed to be second order stationary. The reader may note that stationarity in covariance in the y series is still not guaranteed. If the y series is found to have considerable non-stationarity, perhaps, a suitable transformation on the observed flows prior to harmonic analysis may be tried to render the y series strictly second order stationary. Again the transformations which can be tried are, logarithmic, square root or nth root in the same order. Most often the logarithmic transformation is sufficient. Now the best stationary model is fitted to the y series. First order AR or second order AR model will be the appropriate choice in many cases. The parameters required for the first order AR model are $\bar{y}$, s_y and r_1 denoting, respectively, the mean, the standard deviation and the first serial correlation of the y series. Thus, we see that the number of parameters required for daily flow modelling in this approach is at the maxim 29 or 30 depending on whether, it is the first order AR model or the second order AR model that fits the residual series well.

The parameter to estimated are $\overline{X}$, $\overline{S}$, a_1 to a_6, b_1 to b_6, c_1 to c_6, d_1 to d_6, $\bar{y}$, s_y and r_1, if it is first order AR model and r_2 also if it is second order AR model as both r_1 and r_2 are required in obtaining the two parameters of the second order AR model as well as its multiple correlation co-efficient.

The synthetic daily flow series can now be generated by using the procedure which is the reverse of the analysis described hitherto. First the z_t series is produced using an appropriate random number generation technique such that z_t is from $N(0, 1)$. The y series are now generated from the identified AR model. For example, if the first order AR model is the right choice.

$$y_t = \bar{y} + r_1\,(y_{t-1} - \bar{y}) + z_t . s_y \sqrt{(1 - r_1^2)} \qquad ..(11.73)$$

Then the cyclicity is introduced to get back the daily flows from the equation.

$$x\,(i, j) = y_t .\, s_j + \bar{x}_j \text{ with } t = (i - 1)\,365 + j \qquad ...(11.74)$$

The flows obtained from Eq. (11.74) are to be detransformed if any transformation was used in the analysis. The above modelling procedure can be applied to other seasonal flows such as weekly, or fortnightly flows. In the case of monthly flows the Thomas-Fiering model is preferable because there is not much of difference in the number of parameters required

The modelling of daily flows presented here suffers from one drawback in that it cannot incorporate the characteristic hydrograph shape of a relatively rapid rise followed by a relatively slow exponential decay. Other models such as *shot noise* models avoid this difficulty. The shot noise models characterise the flood peaks by the statistical distribution of times between flood peaks, the statistical distribution of flood magnitudes, and the rate of decline of discharge from each peak value. These models have not yet been extensively used in hydrological investigations. However, if the interest is mainly in the daily discharge volumes, the present model serves the purpose adequately.

EXERCISES

11.1. The annual run off of a river is modelled by $AR(1)$ model with a mean of 8000 Mm^3, standard deviation of 2000 Mm^3 and first serial correlation coefficient of 0.60. Assuming that the annual flows are normally distributed generate a synthetic flow sequence for 10 years given the chain of standard normal random numbers as 0.3138, – 0.3421, – 0.0275, 1.9633, – 1.4020, – 0.5309, 0.7666, – 1.1781, 0.4550 and – 0.0959. What does the generation of negative flows signify ?

11.2. Explain the modifications to be applied to the random component when the annual flows are gamma distributed.

11.3. How do you implement the AR (1) model to generate annual flows when the flows are lognormally distributed ?

11.4. Write down the Thomas-Fiering model for generation of seasonal flows. Explain the significance of each term in the model and explain clearly how the non-stationarity and cyclicity of the seasonal flows are preserved.

11.5. How do you justify the choice of AR models to represent the hydrologic stochastic processes ?

11.6. Write a flow chart and FORTRAN program to generate monthly flows by Thomas-Fiering model.

11.7. Compute the mean, standard deviation and the first serial correlation coefficient of the generated flows of Exercise 11.1 and compare them with the corresponding historical parameters given there.

11.8. The $\bar{x}$, s_x, r_1 and r_2 of observed annual flows of a stream are estimated as 875 Mm^3, 262 Mm^3, 0.8 and 0.46 respectively. Generate a sequence of 3 annual flows assuming that the flows are normally distributed and taking the chain of three standard normal random numbers as 1.265, – 1.006 and 0.091. Use AR (2) model.

11.9. Assuming an $ARMA(1, 1)$ model is a good fit to discribe the normally distributed annual flows of a stream whose mean, standard deviation, first and second serial correlation coefficients are estimated to be 940 Mm^3, 315 Mm^2, 0.56 and 0.44 respectively, generate 3 annual flows. The sequence of 3 standard normal random numbers may be taken as 1.077, 1.027, 0.678.

11.10. Explain what do you understand by Hurst's coefficient and Hurst phenomenon.

11.11. Construct the theoretical correlogram of a *fGn* process with $h = 0.73$, *AR* (1) process with $\rho_1 = 0.37$ and *ARMA* (1, 1) process with $\rho_1 = 0.37$ and $\alpha = 0.9$ on the same graph and comment.

11.12. What is the difference between type II approximation to *fGn*, filtered type II *fGn* and Fast *fGn* models.

11.13. Describe various steps involved in the analysis and modelling of daily flows.

12

Multivariate Stochastic Models

12.1. INTRODUCTION

The models discussed in the previous chapter are concerned with preserving the serial correlation structure and other important statistical properties. Such models may be employed to provide synthetic sequence of annual flows at a single site, for instance, which are required to decide the size and the operation of the reservoir. But, in general, the planning and design of complex water resource systems is concerned with determining the size and operation of not one but several reservoirs at several sites. Thus, quite often simultaneous generation of the synthetic flows at different sites in the same basin may be required. Or sometimes the simultaneous generation of streamflow, sediment discharge and some other water quality variable, all at the same site may be required. Or the generation of streamflow and the cropwater requirement at the same location may be required. In such cases we must consider the models which will preserve not only the serial correlation structure for each variable but also the cross correlation structure between all pairs of variables. For example, when a planning involves a system of sites such as a system of reservoirs it is not satisfactory to simply use single site generation procedure for each of the sites in turn because the flows at various sites may be strongly correlated. If a particular site experiences very large discharge in any month it is very likely that the nearby sites also may experience large discharge in the same month. Moreover, if the flow is large at a particular site during some time period, it will tend to be high some time later at other sites which are down stream in the lower reaches of the stream. Independent generation of the flow values for multiple sites cannot preserve spatial and temporal correlations between flows. Consequently, multisite models, which satisfy these conditions are needed.

In principle, the generation of multivariate synthetic flows is the same as the generation of univariate synthetic flows. A set of flow characteristics to be preserved is specified and a model, constructed to preserve the flow characteristics, is used to generate at each site several synthetic flow sequences each of length equal to the economic time horizon of the system under consideration. In the multivariate case, the set of flow characteristics include those that describe the flow at each site and season of interest, as well as the structure of temporal and spatial interdependence.

12.2. MULTISITE ANNUAL MODELS

Fiering was the first to initiate the work on multisite modelling in hydrology. He proposed what is known as a "principal component" model for multivariate time series. It was later shown by Matalas that the above approach cannot preserve the required correlation structure. Matalas proposed a lag-one multivariate model which is the extension of the first order *AR* model for a univariate case. Since more than one variable are to be treated simultaneously, we require the use of matrix notation in the development of multisite models.

Multisite First Order AR Model. Let $Q(i, t)$ be the flow at the ith site in the year t, where $i = 1, 2, \ldots\ldots, p$ and $t = 1, 2, \ldots\ldots, n$; p and n being the number of sites and number of years of record, respectively. As we are considering the annual flows, the flow series at each site may be assumed to be second order stationary. The mean and standard deviation $\overline{Q}_i$ and s_i of the flows at each site i are computed from

$$\overline{Q}_i = \frac{1}{n} \sum_{t=1}^{n} Q(i, t) \qquad \ldots(12.1)$$

$$s_i = \sqrt{\frac{1}{n} \sum_{t=1}^{n} [Q(i, t) - \overline{Q}_i]^2}$$

$$i = 1, 2, \ldots\ldots, p \qquad \ldots(12.2)$$

Now the annual flow series at all the sites are standardised to have zero mean and unit variance. If $x(i, t)$ represents the standardised flow at site i, and in the year t, then we have

$$x(i, t) = \frac{Q(i, t) - \overline{Q}_i}{s_i} \qquad \ldots(12.3)$$

The matrix of standardised flows represents a multivariate stationary series. The lag-one multivariate model to fit this series is given by

$$\mathbf{X}_t = \mathbf{A}\mathbf{X}_{t-1} + \mathbf{B}\mathbf{Z}_t \qquad \ldots(12.4)$$

where $\mathbf{X}_t$ and $\mathbf{X}_{t-1}$ are p-dimensioned column vectors representing the standardised flows corresponding to p sites at time steps t and $(t-1)$, $\mathbf{Z}_t$ is a column vector of p serially and mutually uncorrelated independent variables at time t with zero mean and unit variance, and **A** and **B** are co-efficient matrices of size (p by p). That is

$$\mathbf{X}_t = \begin{bmatrix} x(1,t) \\ x(2,t) \\ \vdots \\ x(i,t) \\ \vdots \\ x(p,t) \end{bmatrix}, \mathbf{X}_{t-1} = \begin{bmatrix} x(1,t-1) \\ x(2,t-1) \\ \vdots \\ x(i,t-1) \\ \vdots \\ x(p,t-1) \end{bmatrix} \text{ and } \mathbf{Z}_t = \begin{bmatrix} z(1,t) \\ z(2,t) \\ \vdots \\ z(i,t) \\ \vdots \\ z(p,t) \end{bmatrix}$$

Typically, if $a_{i,j}$ and $b_{i,j}$ denote the (i, j)th elements of the matrices **A** and **B**, and if **B** is assumed lower triangular, then

$$x(i,t) = \sum_{j=1}^{p} a_{i,j}\, x(j,t-1) + \sum_{j=1}^{i} b_{i,j}\, z(i,t) \qquad ...(12.5)$$

It may be observed from Eq. (12.5) that the annual flow at any site i at time step t is related to the flow lagged by one time unit at all the sites including the ith site.

The coefficient matrices **A** and **B** are to be determined from the observed data following the procedure explained here. Consider the product of the matrices $\mathbf{X}_t$ and $\mathbf{X}_t^T$, where superscript T stands for the transpose. That is,

$$\mathbf{X}_t\,\mathbf{X}_t^T = \begin{bmatrix} x(1,t) \\ x(2,t) \\ \vdots \\ x(i,t) \\ \vdots \\ x(p,t) \end{bmatrix} [x(1,t)\,x(2,t)\;......\;x(j,t)\;......\;x(p,t)]$$

$$= \begin{bmatrix} x(1,t)\,x(1,t) & x(1,t)\,x(2,t) \ldots & x(1,t)\,x(j,t) \ldots & x(1,t)\,x(p,t) \\ x(2,t)\,x(1,t) & x(2,t)\,x(2,t) \ldots & x(2,t)\,x(j,t) \ldots & x(2,t)\,x(p,t) \\ \vdots & \vdots & \vdots & \vdots \\ x(i,t)\,x(1,t) & x(i,t)\,x(2,t) & x(i,t)\,x(j,t) & x(i,t)\,x(p,t) \\ \vdots & \vdots & \vdots & \vdots \\ x(p,t)\,x(1,t) & x(p,t)\,x(2,t) & x(p,t)\,x(j,t) & x(p,t)\,x(p,t) \end{bmatrix}$$

The expectation of the matrix $\mathbf{X}_t\mathbf{X}_t^T$ is denoted by $\mathbf{M}_0$. Therefore

$$\mathbf{M}_0 = E\,[\mathbf{X}_t\,\mathbf{X}_t^T] \qquad \text{...(12.6)}$$

If $m_0\,(i, j)$ is the general element of $\mathbf{M}_0$ matrix in the ith row and jth column, then it is given by

$$m_0\,(i, j) = \frac{1}{n}\,[x(i, 1)\,x(j, 1) + x(i, 2)\,x(j, 2) + \ldots\ldots + x\,(i, n)\,x(j, n)]$$

$$= \frac{1}{n} \sum_{t=1}^{n} x(i, t)\,x(j, t)$$

$$= \frac{1}{n} \sum_{t=1}^{n} \left[\left\{ \frac{Q(i, t) - \overline{Q}_i}{s_i} \right\} \left\{ \frac{Q(j, t) - \overline{Q}_j}{s_j} \right\} \right]$$

Thus we see that $m_0\,(i, j)$ is nothing but the correlation coefficient between the flows at sites i and j during the same time period. Also when $i = j$, $m_0\,(i, j) = 1$. Therefore, the matrix $\mathbf{M}_0$ is called *cross covariance matrix* of lag zero and it is a symmetric matrix with all its diagonal elements as 1.

Similarly consider the product of the matrices $\mathbf{X}_t$ and $\mathbf{X}^T_{t-1}$ and denote the expectation of this matrix by $\mathbf{M}_1$. That is

$$\mathbf{M}_1 = E\,[\mathbf{X}_t\,\mathbf{X}^T_{t-1}] \qquad \text{...(12.7)}$$

We have

$$\mathbf{X}_t\,\mathbf{X}^T_{t-1} = \begin{bmatrix} x(1, t) \\ x(2, t) \\ \vdots \\ x(i, t) \\ \vdots \\ x(p, t) \end{bmatrix} [x(1, t-1), x(2, t-1) \ldots\ldots x(j, t-1) \ldots\ldots x(p, t-1)]$$

$$= \begin{bmatrix} x(1, t)\,x(1, t-1) & x(1, t)\,x(2, t-1) \ldots & x(1, t)\,x(j, t-1) \ldots & x(1, t)\,x(p, t-1) \\ x(2, t)\,x(1, t-1) & x(2, t)\,x(2, t-1) \ldots & x(2, t)\,x(j, t-1) \ldots & x(2, t)\,x(p, t-1) \\ \vdots & & & \\ x(i, t)\,x(1, t-1) & x(i, t)\,x\,(2, t-1) \ldots & x\,(i, t)\,x\,(j, t-1) \ldots & x(i, t)\,x(p, t-1) \\ \vdots & & & \\ x(p, t)\,x(1, t-1) & x(p, t)\,x(2, t-1) \ldots & x(p, t)\,x(j, t-1) \ldots & x(p, t)\,x(p, t-1) \end{bmatrix}$$

If $m_1\,(i, j)$ is the general element of $\mathbf{M}_1$ matrix in the ith row and jth column, then it is given by

$$m_1(i, j) = \frac{1}{n}[x(i, 1)\, x(j, 0) + x(i, 2)\, x(j, 1) + \ldots\ldots + x(i, n)\, x(j, n-1)]$$

Since the record is available only for $t = 1$ onwards, we can write

$$m_1(i, j) = \frac{1}{(n-1)} \sum_{t=2}^{n} [x(i, t)\, x(j, t-1)]$$

$$m_1(i, j) = \frac{1}{(n-1)} \sum_{t=2}^{n} \left\{ \frac{Q(i, t) - \overline{Q}_i}{s_i} \right\} \left\{ \frac{Q(j, t-1) - \overline{Q}_j}{s_j} \right\}$$

or $\quad m_1(i, j) = r_{i,j}(-1)$ ...(12.8)

Thus we see that $m_1(i, j)$ represents $r_{i,j}(-1)$ which is the lag-one cross correlation between the flows at ith and jth sites. Here the (-1) in $r_{i,j}(-1)$ signifies that in the cross correlation the flow series at the jth site is lagged behind the flow series at the ith site by one time unit when computing the sum of the cross products. Likewise, the elements of the matrix which is defined as $E[\mathbf{X}_{t-1}\, \mathbf{X}_t^T]$ will denote the cross correlation $r_{i,j}(+1)$. It may be noted that $r_{i,j}(-1) \neq r_{i,j}(+1)$. For diagonal elements of $\mathbf{M}_1$ matrix where $i = j$, $r_{i,j}(-1)$ represents the lag-one serial correlation coefficient of the flows at site i and in this case if the flow at site i are assumed to be stationary then $r_{i,j}(-1) = r_{i,j}(+1)$. The $\mathbf{M}_1$ matrix is called the *cross covariance matrix of lag one* and, unlike $\mathbf{M}_0$, it is not symmetric.

The co-efficient matrices **A** and **B** can now be obtained in terms of the covariance matrices $\mathbf{M}_0$ and $\mathbf{M}_1$. Post multiplying Eq. (12.4) on both sides by $\mathbf{X}_{t-1}^T$ and taking expectations, we get

$$E[\mathbf{X}_t\, \mathbf{X}^T_{t-1}] = \mathbf{A}.E[\mathbf{X}_{t-1}\, \mathbf{X}^T_{t-1}] + \mathbf{B}.E[\mathbf{Z}_t\, \mathbf{X}^T_{t-1}]$$

Since $\mathbf{Z}_t$ and $\mathbf{X}_{t-1}$ are independent of each other, it leads to

$$\mathbf{M}_1 = \mathbf{A}.\mathbf{M}_0 + 0$$

or $\quad \mathbf{A} = \mathbf{M}_1 \,.\, \mathbf{M}_0^{-1}$ ...(12.9)

where the superscript -1 stands for the inverse of the matrix. Post multiplying Eq. (12.4) on both sides by $\mathbf{X}_t^T$ and taking expectations.

$$E[\mathbf{X}_t\, \mathbf{X}_t^T] = \mathbf{A}.E[\mathbf{X}_{t-1}\, \mathbf{X}_t^T] + \mathbf{B}.E[\mathbf{Z}_t \,.\, \mathbf{X}_t^T] \quad \ldots(12.10)$$

While the left hand side of the above equation is equal to $\mathbf{M}_0$, the two terms on the right hand side can be simplified as follows. We have by definition.

$$\mathbf{M}_1 = E[\mathbf{X}_t \,.\, \mathbf{X}^T_{t-1}]$$

Transposing the above equation on both sides

$$\mathbf{M}_1^T = \{E[\mathbf{X}_t \,.\, \mathbf{X}^T_{t-1}]\}^T$$
$$= E[\{\mathbf{X}_t \,.\, \mathbf{X}^T_{t-1}\}^T]$$
$$\mathbf{M}_1^T = E[\mathbf{X}_{t-1}\, \mathbf{X}_t^T] \qquad ...(12.11)$$

Now, $\mathbf{Z}_t \,.\, \mathbf{X}_t^T = \mathbf{Z}_t\, [\mathbf{A} \,.\, \mathbf{X}_{t-1} + \mathbf{B}.\mathbf{Z}_t]^T$

$$= \mathbf{Z}_t \,.\, \mathbf{X}^T_{t-1}\, \mathbf{A}^T + \mathbf{Z}_t \,.\, \mathbf{Z}_t^T\, \mathbf{B}^T$$

Now taking the expectations on both sides, we have

$$E[\mathbf{Z}_t \,.\, \mathbf{X}_t^T] = O + \mathbf{I}.\mathbf{B}^T = \mathbf{B}^T \qquad ...(12.12)$$

where $\mathbf{I}$ is the identity matrix. The above result is due to the fact that $\mathbf{Z}_t$ and $\mathbf{X}_{t-1}$ are independent and $\mathbf{Z}_t$ is an independent process with unit variance. Substituting the results of Eq. (12.11) and Eq. (12.12) in Eq. (12.10), we get

$$\mathbf{M}_0 = \mathbf{A}.\mathbf{M}_1^T + \mathbf{B}.\mathbf{B}^T$$

or $$\mathbf{B}\mathbf{B}^T = \mathbf{M}_0 - \mathbf{M}_1\mathbf{M}_0^{-1}\,\mathbf{M}_1^T \qquad ...(12.13)$$

since $\mathbf{A} = \mathbf{M}_1.\mathbf{M}_0^{-1}$ from Eq. (12.9). The co-efficient matrices $\mathbf{A}$ and $\mathbf{B}$ can be obtained by solving Eq. (12.9) and (12.13).

If $$\mathbf{B}\mathbf{B}^T = \mathbf{C} \qquad ...(12.14)$$

then $$\mathbf{C} = \mathbf{M}_0 - \mathbf{M}_1\mathbf{M}_0^{-1}\,\mathbf{M}_1^T \qquad ...(12.15)$$

from which $\mathbf{C}$ can be estimated by substituting the estimated values of the elements of matrices $\mathbf{M}_0$ and $\mathbf{M}_1$. It can be easily verified that $\mathbf{C}$ is necessarily symmetric. In contrast with matrix $\mathbf{A}$, matrix $\mathbf{B}$ does not have a unique solution. In fact there are infinite number of solutions for $\mathbf{B}$ which satisfy Eq. (12.14). One method which is comparatively straight forward is to assume that $\mathbf{B}$ is lower triangular. Accordingly $\mathbf{B}\mathbf{B}^T$ and $\mathbf{C}$ are written as follows.

$$\mathbf{B}\mathbf{B}^T = \begin{bmatrix} b_{1,1} & 0 & 0 & 0 \ldots 0 \\ b_{2,1} & b_{2,2} & 0 & 0 \ldots 0 \\ b_{3,1} & b_{3,2} & b_{3,3} & 0 \ldots. 0 \\ \vdots & \vdots & \vdots & \\ b_{p,1} & b_{p,2} & b_{p,3} & \ldots b_{p,p} \end{bmatrix} \begin{bmatrix} b_{1,1} & b_{2,1} & b_{3,1} \cdots & b_{p,1} \\ 0 & b_{2,2} & b_{3,2} \cdots & p_{p,2} \\ 0 & 0 & b_{3,3} \cdots & b_{p,3} \\ \vdots & \vdots & \vdots & \vdots \\ 0 & 0 & 0 & b_{p,p} \end{bmatrix}$$

$$\mathbf{C} = \begin{bmatrix} c_{1,1} & c_{1,2} & c_{1,3} \cdots\cdots & c_{1,p} \\ c_{2,1} & c_{2,2} & c_{2,3} \cdots\cdots & c_{2,p} \\ \vdots & \vdots & \vdots & \vdots \\ c_{p,1} & c_{p,2} & c_{p,3} \cdots\cdots & c_{p,p} \end{bmatrix}$$

The diagonal elements are found from

$$b_{1,1} = (c_{1,1})^{1/2}$$

$$b_{2,2} = (c_{2,2} - b^2_{2,1})^{1/2}$$

$$\vdots \qquad \vdots$$

In general,

$$b_{k,k} = (c_{k,k} - b^2_{k,k-1} - b^2_{k,k-2} - \dots\dots - b^2_{k,1}) \qquad \dots(12.16)$$

For these relations to hold, **C** should be positive definite ; thus the elements of **C** are constrained as follows.

$$c_{1,1} \geq 0$$

$$c_{2,2} \geq b^2_{2,1}$$

$$c_{3,3} \geq b^2_{3,2} + b^2_{3,1}$$

$$\vdots \qquad \vdots \qquad \vdots$$

$$c_{k,k} \geq b^2_{k,k-1} + b^2_{k,k-2} + \dots\dots + b^2_{k,1} \qquad \dots(12.17)$$

Also it follows from Eq. (12.14) that, the elements in the kth row are given by

$$b_{k,1} = c_{k,1}/b_{1,1}$$

$$b_{k,2} = (c_{k,2} - b_{2,1}\, b_{k,1})/b_{2,2}$$

$$b_{k,3} = (c_{k,3} - b_{3,1}\, b_{k,1} - b_{3,2}\, b_{k,2})/b_{3,3}$$

$$\vdots \qquad \vdots \qquad \vdots \qquad \vdots$$

$$b_{k,j} = (c_{k,j} - b_{j,1}\, b_{k,1} - b_{j,2}\, b_{k,2} - \dots\dots - b_{j,j-1}\, b_{k,j-1})/b_{j,j} \qquad \dots(12.18)$$

Solving Eq. (12.16) and Eq. (12.18) the elements of **B** matrix are obtained.

In order to fit the lag-one multivariate *AR* (1) model described above to any data both $\mathbf{M}_0$ and $\mathbf{BB}^T$ should be positive semi-definite. Which means that the determinant values of $\mathbf{M}_0$ and $\mathbf{BB}^T$ should be greater than zero. This condition may be utlised to check the inconsistency in the data, if any.

In the above analysis there is an implicit assumption that the model is multivariate normal. That means, the flows at each site are assumed to be normally distributed. In case the flows are not normally distributed and the skewness is an important parameter to be preserved, the analysis becomes exceedingly complex.

Once the co-efficient matrices are estimated from the observed data, the generation procedure follows the steps in the reverse order. If the generation of multisite flows is required for N years, we generate the sequence of the normal random deviate z of length $N.p$, where p is the number of sites. Taking p random number each time we construct $\mathbf{Z}_t$ column vectors. That is

$$\mathbf{Z}_1 = [z_1, z_2, \ldots\ldots, z_p]^T$$
$$\mathbf{Z}_2 = [z_{p+1}, z_{p+2}, \ldots\ldots, z_{2p}]^T \text{ etc.}$$

To generate the standardised flows $x(i, 1)$ at first time step we require $x(i, 0)$ which are the standardised flows corresponding to the previous time step. As in the case of univariate models, the column vector $\mathbf{X}_0$ becomes the seed of generation in the multivariate case. Since the standardised flows of all the sites have the mean equal to zero, $\mathbf{X}_0$ is taken to be a column vector with all its elements as zero. That means

$$\mathbf{X}_0 = [0, 0, \ldots\ldots, 0]^T.$$

Now $\mathbf{X}_1$ is generated from

$$\mathbf{X}_1 = \mathbf{A}.\mathbf{X}_0 + \mathbf{B}.\mathbf{Z}_1.$$

Once $\mathbf{X}_1$ is obtained the subsequent column matrices are generated in interative steps till, we get $\mathbf{X}_N$. Now the flows are obtained by introducing necessary mean and variance

$$Q(i, t) = s_i x(i, t) + \overline{Q}_i \qquad \ldots(12.19)$$

with $i = 1, 2, \ldots\ldots p$ and
$t = 1, 2, \ldots\ldots N.$

Example 12.1. *The following cross correlation matrices are estimated from the historical data of annual streamflows at two sites*

$$\mathbf{M}_0 = \begin{bmatrix} 1 & 0.8547 \\ 0.8547 & 1 \end{bmatrix}, \mathbf{M}_1 = \begin{bmatrix} 0.6797 & 0.5374 \\ 0.4635 & 0.4939 \end{bmatrix}$$

At site 1 the mean annual flow is 240 Mm^3 and the standard deviation is 147 Mm^3 and at site 2 they are 79 Mm^3 and 46 Mm^3 respectively. Generate the first two values of the data for these two sites using the following random numbers for the elements in the vector $\mathbf{Z}_t$ – 0.134, – 0.268, 1.639, 0.134.

Sol. Let d be the determinant of the matrix $\mathbf{M}_0$. Then

$d = 1 - 0.8547^2 = 0.2695$. And now

$$\mathbf{M}_0^{-1} = \begin{bmatrix} \dfrac{1}{d} & \dfrac{-0.8547}{d} \\ \dfrac{-8547}{d} & \dfrac{1}{d} \end{bmatrix} = \begin{bmatrix} 3.7106 & -3.1714 \\ -3.1714 & 3.7106 \end{bmatrix}$$

$$\mathbf{A} = \mathbf{M}_1 \mathbf{M}_0^{-1} = \begin{bmatrix} 0.6797 & 0.5374 \\ 0.4635 & 0.4939 \end{bmatrix} \begin{bmatrix} 3.7106 & -3.1714 \\ -3.1714 & 3.7106 \end{bmatrix}$$

$$= \begin{bmatrix} 0.8178 & -0.1615 \\ 0.1535 & 0.3627 \end{bmatrix}$$

$$\mathbf{BB}^T = \mathbf{C} = \mathbf{M}_0 - \mathbf{M}_1 \mathbf{M}_0^{-1} \mathbf{M}_1^T$$

$$= \begin{bmatrix} 1 & 0.8547 \\ 0.8547 & 1 \end{bmatrix} - \begin{bmatrix} 0.8178 & -0.1615 \\ 0.1535 & 0.3627 \end{bmatrix} \begin{bmatrix} 0.6797 & 0.4635 \\ 0.5374 & 0.4939 \end{bmatrix}$$

$$= \begin{bmatrix} 1 & 0.8547 \\ 0.8547 & 1 \end{bmatrix} - \begin{bmatrix} 0.4691 & 0.2993 \\ 0.2993 & 0.2503 \end{bmatrix}$$

$$= \begin{bmatrix} 0.5309 & 0.5554 \\ 0.5554 & 0.7497 \end{bmatrix}$$

$$\mathbf{C} = \begin{bmatrix} 0.5309 & 0.5554 \\ 0.5554 & 0.7497 \end{bmatrix}$$

Using Eqs. (12.16) and (12.18)

$$b_{1,1} = (c_{1,1})^{1/2} = (0.5309)^{1/2} = 0.729$$

$$b_{2,1} = \frac{c_{2,1}}{b_{1,1}} = \frac{0.5554}{0.729} = 0.762$$

$$b_{2,2} = (c_{2,2} - b^2_{2,1})^{1/2} = (0.7497 - 0.762^2)^{1/2} = 0.411$$

$$\mathbf{B} = \begin{bmatrix} 0.729 & 0 \\ 0.762 & 0.411 \end{bmatrix}$$

$$\mathbf{X}_t = \mathbf{AX}_{t-1} + \mathbf{BZ}_t$$

$$\begin{bmatrix} x_{1,t} \\ x_{2,t} \end{bmatrix} = \begin{bmatrix} 0.8178 & -0.1615 \\ 0.1535 & 0.3627 \end{bmatrix} \begin{bmatrix} x_{1,t-1} \\ x_{2,t-1} \end{bmatrix} + \begin{bmatrix} 0.729 & 0 \\ 0.762 & 0.411 \end{bmatrix} \begin{bmatrix} z_{1,t} \\ z_{2,t} \end{bmatrix}$$

$$\begin{bmatrix} x_{1,1} \\ x_{2,1} \end{bmatrix} = \begin{bmatrix} 0.8178 & -0.1615 \\ 0.1535 & 0.3627 \end{bmatrix} \begin{bmatrix} x_{1,0} \\ x_{2,0} \end{bmatrix} + \begin{bmatrix} 0.729 & 0 \\ 0.762 & 0.411 \end{bmatrix} \begin{bmatrix} -0.134 \\ -0.268 \end{bmatrix}$$

where $x_{1,0}$ and $x_{2,0}$ are the seeds which are assumed to be zero.

$$\text{Then} \begin{bmatrix} x_{1,1} \\ x_{2,1} \end{bmatrix} = \begin{bmatrix} -0.0977 \\ -0.2122 \end{bmatrix}$$

$$\begin{bmatrix} x_{1,2} \\ x_{2,2} \end{bmatrix} = \begin{bmatrix} 0.8178 & -0.1615 \\ 0.1535 & 0.3627 \end{bmatrix} \begin{bmatrix} -0.0977 \\ -0.2122 \end{bmatrix} + \begin{bmatrix} 0.729 & 0 \\ 0.762 & 0.411 \end{bmatrix} \begin{bmatrix} 1.639 \\ 0.134 \end{bmatrix}$$

$$= \begin{bmatrix} -0.0456 \\ -0.0920 \end{bmatrix} + \begin{bmatrix} 1.1948 \\ 1.3040 \end{bmatrix} = \begin{bmatrix} 1.1492 \\ 1.2120 \end{bmatrix}$$

At site 1,

$$Q_{1,1} = \overline{Q}_1 + x_{1,1}\, s_1 = 240 - 0.0977 \times 147 = 225.6 \text{ Mm}^3$$

$$Q_{1,2} = \overline{Q}_1 + x_{1,2}\, s_1 = 240 + 1.1492 \times 147 = 408.9 \text{ Mm}^3$$

At site 2,

$$Q_{2,1} = \overline{Q}_2 + x_{2,1}\, s_2 = 79 - 0.2122 \times 46 = 69.2 \text{ Mm}^2$$

$$Q_{2,2} = \overline{Q}_2 + x_{2,2}\, s_2 = 79 + 1.212 \times 46 = 134.8 \text{ Mm}^3.$$

Multisite higher order AR model. The first order *AR* model of Eq. (12.4) can be extended to form a general higher order multisite *AR* model as follows

$$\mathbf{X}_t = \sum_{k=1}^{m} \mathbf{A}_k \mathbf{X}_{t-k} + \mathbf{B}\mathbf{Z}_t \qquad \text{...(12.20)}$$

where m is the order of the model. Let $\mathbf{M}_0$ denote $(p \times p)$ lag-zero cross covariance matrix and let $\mathbf{M}_k$, for $k = 1, 2, \ldots\ldots, m$ denote $(p \times p)$ cross covariance matrix of lag k. Here, the materix $\mathbf{M}_0$ is as defined in Eq. (12.6). However, $\mathbf{M}_k$ is such that its diagonal elements are the lag-k serial correlation coefficients at each of the p sites in order and its general (i, j)th element represents $r_{i,j}(-k)$ which is the lag k cross correlation coefficient between the flows at ith and jth sites with the flows at the jth site lagged behind the flows at the ith site by k time units. That is

$$\mathbf{M}_k = E\left[\mathbf{X}_t\, \mathbf{X}^T_{t-k}\right] \qquad \text{...(12.21)}$$

If $m_k(i, j)$ is the general element of $\mathbf{M}_k$ in the ith row and jth column, then it is given by

$$m_k(i, j) = \frac{1}{n}\,[x(i, 1)\, x(j, 1-k) + x(i, 2)\, x(j, 2-k) + \ldots\ldots + x(i, n)\, x(j, n-k)]$$

Since the record is available only for $t = 1$ onwards, we can write

$$m_k(i, j) = \frac{1}{(n-k)} \sum_{t=k+1}^{n} x(i, t)\, x(j, t-k) \qquad \text{...(12.22)}$$

Then the matrics $\mathbf{B}$ and $\mathbf{A}_k$, $k = 1, 2, \ldots\ldots, m$ can be obtained by making use of the relations that exist between these matrices and $\mathbf{M}_0$ and $\mathbf{M}_k$, $k = 1, 2, \ldots\ldots, m$.

12.3. MULTISITE AR MODEL FOR SEASONAL FLOWS

The lag-one multivariate model described above is essentially stationary and therefore it cannot be applied directly to multisite seasonal flows since the seasonal flows always exhibit non-stationary behaviour because of inherent cyclic component present in them. Young and Pisano have applied the lag-one multivariate model for monthly flows taking the logarithmic transformation and making them approximately stationary. Their procedure is described below.

Let $Q(k, i, j)$ be the flow in jth month of ith year at site k where $k = 1, 2, \ldots\ldots p$; $j = 1, 2, \ldots\ldots, 12$; and $i = 1, 2, \ldots\ldots, n$. n being the number of years of record. Now a suitable transformation like logarithmic or square root which ever gives minimum skewness co-efficient is applied. After their transformation the flows in all seasons are assumed to be normally distributed as well as stationary in covariance. Usually the logarithmic transformation is sufficient. Therefore

$$q(k, i, j) = \ln [Q(k, i, j)] \qquad ..(12.23)$$

Now, to make the q series stationary in mean and variance also, the cyclicity in mean and variance are removed through the following transformation

$$\bar{q}(k, j) = \frac{1}{n} \sum_{i=1}^{n} q(k, i, j) \qquad ...(12.24)$$

$$s(k, j) = \sqrt{\frac{1}{n} \sum_{i=1}^{n} [q(k, i, j) - \bar{q}(k, j)]^2} \qquad ...(12.25)$$

$$x(k, t) = \frac{q(k, i, j) - \bar{q}(k, j)}{s(k, j)} \qquad ...(12.26)$$

where $t = (i - 1)\, 12 + j$. The multivariate series 'x' are now supposed to be second order stationary and a lag-one multivariate model is fitted to these series following the same procedure of annual multisite model. Again, in the generation of multisite monthly flows the reverse procedure is followed. First the 'x' series are generated using the lag-one multivariate model. Then the cyclicity in the mean and standard deviation are reintroduced to get the logarithms of the flows. Finally by taking antilogarithms the flows are obtained. The various steps are as follows :

$$\mathbf{X}_t = \mathbf{A}\mathbf{X}_{t-1} + \mathbf{B}\mathbf{Z}_t$$

$$q(k, i, j) = x(k, t) \,.\, s(k, j) + \bar{q}(k, j) \qquad ...(12.27)$$

where $t = (i - 1)\, 12 + j$

$$Q(k, i, j) = e^{[q(k, i, j)]} \qquad ...(12.28)$$

The above approach suggested by Young and Pisano is very attractive. It requires very few parameters to be estimated. But it is based on restrictive assumption that the residuals are stationary after normalising (logarithmic) and standardising (removal of cyclicity) transformations, which may not be true always. Nevertheless, this approach has been used successfully in many cases. Moreover when the data is short it becomes difficult to apply the non-stationary multivariate model which is described below.

Non-Stationary Multivariate Seasonal Model. A non-stationary multivariate model may be fitted to multisite monthly data on the same lines as Thomas-Fiering model for a single site. We may use the lag-one multivariate model. The only difference is that the co-efficient matrices will no longer be constant but they will vary from month to month. The model is written in the form

$$\mathbf{X}_t = \mathbf{A}_j \mathbf{X}_{t-1} + \mathbf{B}_j \mathbf{Z}_t \quad \text{...(12.29)}$$

where $j = t \bmod (12)$, and $j = 12$ if $t \bmod (12)$ is zero, $\mathbf{A}_j$ and $\mathbf{B}_j$ are the coefficient matrices for the month j which are estimated from

$$\mathbf{A}_j = \mathbf{M}_{1,j} \mathbf{M}^{-1}_{0,j} \quad \text{...(12.30)}$$

$$\mathbf{B}_j \mathbf{B}_j^T = \mathbf{M}_{0,j} - \mathbf{M}_{1,j} \mathbf{M}^{-1}_{0,j} \mathbf{M}^T_{1,j} \quad \text{...(12.31)}$$

where $\mathbf{M}_{0,j}$ and $\mathbf{M}_{1,j}$ are defined in a similar fashion to that of $\mathbf{M}_0$ and $\mathbf{M}_1$ matrices of multivariate first order *AR* model except that when evaluating its elements the summation will be running with a jump of 12 each time. However, when the available record is short, the estimation of these coefficient matrices becomes unreliable since there will be fewer terms while taking expectations. Again the condition that the determinant value of $\mathbf{M}_{0,i}$ and $\mathbf{B}_j \mathbf{B}_j^T$ should be greater than zero must be ascertained for each month j.

12.4. OTHER TYPES OF MULTISITE MODELS

In the recent investigations, multisite modelling is attempted using the ARIMA models and *fGn* models. Those details are, however, not presented here since the subject matter is at an advanced level and is beyond the scope of treatment here.

EXERCISES

12.1. Why multivariate models are needed in hydrology ? How are they different from univariate models.

12.2. Write down the multisite *AR*(1) model for annual flows and explain clearly how the matrices $\mathbf{M}_0$ and $\mathbf{M}_1$ are defined and how the coefficient matrices **A** and **B** are estimated.

12.3. The $\mathbf{M}_0$ and $\mathbf{M}_1$ matrices estimated from the observed annual flows at two sites are

$$\begin{bmatrix} 1.00 & 0.86 \\ 0.86 & 1.00 \end{bmatrix} \text{ and } \begin{bmatrix} 0.71 & 0.65 \\ 0.51 & 0.53 \end{bmatrix}$$

The mean annual flow for the two sites are 412 Mm^3 and 936 Mm^3, and the standard deviation of the annual flow at the two sites are 178 Mm^3 and 571 Mm^3 respectively. Generate two flows at each site using the multisite AR (1) model. The standard normal random variates may be takes as 0.147, – 0.797, – 1.221 and 1.416.

12.4. In AR (1) model for multisite annual flows, why all the diagonal elements of $\mathbf{M}_0$ matrix are 1? What do they represent. What do the diagonal elements of $\mathbf{M}_1$ matrix represent? Unlike $\mathbf{M}_0$ matrix why $\mathbf{M}_1$ matrix is not symmetric.

12.5. Explain various steps involved in modelling of multisite seasonal flows by $AR(1)$ model. What are its limitations.

12.6. Write down the multisite AR (1) model for the seasonal flows. Explain how the matrices $\mathbf{M}_{0,j}$ and $\mathbf{M}_{i,j}$ are defined. Give an expression to evaluate a typical element of these matrices in row u and column v, that is $m_{0,j}(u, v)$ and $m_{1,j}(u, v)$.

13

Generation of Random Variates

13.1. INTRODUCTION

As we have seen in the previous chapter, once a proper model is identified and its parameters estimated the first step in the generation of synthetic flow sequences is to produce the independent random series z_t with a specific mean and a specific variance and drawn from a specified distribution. This random series should not exhibit serial correlation for any lag. Fortunately the random numbers of any distribution can be obtained by a transformation from the random numbers of uniform distribution. We shall therefore first discuss the methods of generating uniformly distributed numbers and later show how these can be transformed into random numbers of other distributions. As already described a uniformly distributed variable is the one for which each value in a given range (say 0 to 1) is equally likely.

Strictly speaking random numbers can be obtained only from random processes, since the word random implies that the numbers obtained are unpredictable. Various mechanical, electrical and electronic devices have been built in the past to produce such numbers. These methods include experimentation with numbers drawn from a well-shaken stack of numbered chits, selecting numbers at random from telephone directories, use of an electronic *roulette* wheel etc. However, there are number of disadvantages associated with the use of these methods. They are

(*i*) Nature tends to be systematic

(*ii*) The equipment tends to be cumbersome and difficult to maintain, and

(*iii*) It is impossible to repeat a simulation using the same set of random numbers.

The last point is a very important one. This is because a highly desirable feature of simulation models is that one should be able to repeat the simulation process changing only one exogenous variable and keeping everything else constant. This requires the ability to generate exactly the same sequence of random numbers over and over again. When we are using a computer to generate the flows, we could store a table of random numbers on a disc or tape and use them whenever we want. But they consume considerable core storage depending on the magnitude of the problem and also the retrieval of random numbers from the memory is a delayed process. So, from the view point of simulation on a fast digital computer, the sequence of random numbers produced by any method should satisfy the following conditions, *i.e.*, it should be

(*i*) Capable of being generated at high speed.

(*ii*) Reproducible.

(*iii*) Non-repeating for any specified length (without periodicity).

(*iv*) Statistically independent.

(*v*) Needing a minimum amount of core storage.

(*vi*) Uniformly distributed.

To satisfy the aforestated requirements, many techniques have been suggested which use some form of the recurrence relationship. Since the numbers that are generated by means of an algorithm based on any recurrence relationship can be predicted, they are called *pseudorandom* numbers. If these pseudo-random numbers satisfy the other tests of randomness they can be accepted as random numbers.

13.2. UNIFORMLY DISTRIBUTED RANDOM NUMBERS

An algorithm for generating pseudorandom sequence of numbers is commonly called a *random number generator*. There are many algorithms for generating random numbers, differing from one another in the extent to which they exhibit the desirable properties outlined earlier.

Mid-square technique. One of the earliest techniques for the generation of pseudorandom numbers is the mid-square technique. In this method the kth random number of $2d$ digits is obtained by squaring the $(k-1)$th random number to produce a number of $4d$ digits and by selecting the middle $2d$ digits from the number so produced. To initiate the generation we require a $2d$ digit which is called the *random seed* or simply the *seed*. If the seed u_0 is taken as 5069, for example, then the first random number u_1 is obtained by squaring u_0 and taking the middle 4 digits. Thus

$$u_0^2 = (5069)^2 = 25694761 \text{ and } u_1 = 6947.$$

Proceeding in a similar way, we get the chain as $u_2 = 2608$, $u_3 = 8016$, $u_4 = 2562$, ... and so on. This scheme has certain deficiencies. In the chain if a number 0000 appears, then the scheme terminates. Also if the seed is not selected properly it may lead to very short cycles.

Mid-product technique. Another technique which was tried was the mid-product technique. In this method the kth random number is taken as the central $2d$ digits of the product of $(k - 1)$th and $(k - 2)$th random numbers, and requires two initial numbers each of $2d$ digits to initiate the generation procedure. The mid-product technique also has the same disadvantages as the mid-square technique.

GPSS algorithm. Another commonly used random-number generator having good statistical properties is the GPSS algorithm. The GPSS generator gets its name from *General Purpose System Simulator*. The idea of this algorithm can best be illustrated by an example using decimal integers, although it is actually carried out using binary numbers on a computer. Suppose, for the sake of example, that it is desired to generate random number u_k in the range 0000 to 9999. We begin with a multiplies m, which is fixed and a seed r_0. Let $m = 5167$ and $r_0 = 3729$. These are multiplied together to produce a product having eight digits. That is, $mr_0 = 5167 \times 3729 = 19267743$. The middle 4 digits of this product are used to form the first random number u_1 and the right most 4 digits are used as r_1. Thus $u_1 = 2677$ and $r_1 = 7743$. Next the product of mr_1 is formed, its middle 4 digits are used as the second random number u_2, its right most 4 digits are used as r_2, and so forth. Later a decimal is put is front of each random number to get the sequence of random numbers in the range 0 to 0.9999.

The multiplicative Congruential method. Improvements over the mid-square and mid-product technique were incorporated in the multiplicative congruential method which is based on the recursion formula

$$u_k = (a\, u_{k-1}) \bmod m \qquad ...(13.1)$$

where the parameter a is called the *multiplier*, the parameter m is called the *modulus* and starting number u_0 is called the *seed*. The notation "mod m" in this formula refers to an integer division. That is, x mod m = remainder left over after dividing x by m. Eq. (13.1) can also be written in the form

$$u_k = a^k\, u_0 \bmod m \qquad ...(13.2)$$

For this reason, this method is also known as the *power residue method*. The sequence of random numbers generated by the congruential method will be in the range 0 to $(m - 1)$ and will eventually be repeated. The period of repetition will always be less than m. So m should be chosen as the largest value that can be conveniently used.

If the multiplicative congruential generator is programmed on a binary computer the choice of the various constants should be as follows

$m = 2^b$, where b is the number of bits per word

$a = 8t \pm 3$, t is any integer ≥ 0, and

u_0 = any odd integer.

This choice provides the maximum period of 2^{b-2} random numbers in the range 0 to 2^b. By dividing every number with 2^b, the chain provides the random numbers in the range 0 to 1. When the multiplicative congruential generator is programmed on a decimal computer, the choice is

$m = 10^d$, where d is the number of digits per word

$a = 200t + p$, where t is any integer ≥ 0, and p is a number from the set of 32 integers given by

$\pm \{3, 11, 13, 19, 21, 27, 29, 37, 53, 59, 61, 67, 69, 77, 83, 91\}$

u_0 = any number not divisible by either 2 or 5.

This choice provides a maximum period of $(\frac{1}{2}) \cdot 10^{d-2}$, if $d \geq 4$.

Again, by putting decimal before every number, we get a chain of uniformly distributed random numbers in the range 0 to 1.

Mixed congruential method. This method is also known as *linear congruential method*. The recursion formula in this method is given by

$$u_k = (a\, u_{k-1} + c) \bmod m \qquad \text{...(13.3)}$$

where the number c is called the *increment* and a, m and u_0 retain the same names as in the multiplicative congruential method. Improved statistical properties of the sequence can be obtained in the mixed congruential method compared to multiplicative congruential method.

The multiplicative generator produces a sequence of random numbers whose period is always less than m, whether it is implemented on a binary or decimal computer. Whereas in the mixed congruential method when a, c and m are chosen according to the conditions given below, a sequence of m pseudorandom numbers in the range 0 to $(m - 1)$ can be generated before the sequence begins to repeat.

Binary computer	*Decimal computer*
$m = 2^b$	$m = 10^d$
c = any odd integer	c = any integer not divisible by either 2 or 5
$a \bmod 4 = 1$	$a \bmod 20 = 1$
u_0 = any integer so that $0 \le u_0 \le m$	u_0 = any integer so that $0 \le u_0 \le m$

The selection of a and c in accordance with the above specifications implies that all integers less than m will appear before the seed appears. In order that the sequence of random numbers generated is independent of internal serial correlation, the additional specifications imposed on a and c are

$$a \simeq \sqrt{m}$$

$$c \simeq \frac{m}{2}$$

Finally the uniform random numbers in the range 0 to 1 may be obtained by dividing the sequence by m.

In order to reduce the serial dependence in the series, a second sequence of random numbers using the recursion equation

$$u_k' = (a' \, u_{k-1}' + c') \bmod m'$$

and the seed u_0' is generated and dividing the second sequence by m' it is combined alternately with the first sequence.

Testing the random number sequence. The chi-square test may be performed to ascertain whether the generated sequence of numbers follow uniform distribution or not. Treating the generated sequence as a time series one can construct the correlogram for sufficiently large number of lags and the corresponding spectral density function and test them for independence by obtaining the tolerance limits as per the procedures described elsewhere in the text. Experience has shown that the sequence of random numbers generated by mixed congruential method passes most of the tests for independence.

Example 13.1. *Generate a sequence of 12 uniformly distributed random numbers of 5 digits using the mixed congruential technique. Choose the appropriate values for the constants a, c and m.*

Sol. In accordance with the stipulations on the choice of the parameters given for mixed congruential method, the parameters are chosen as given below.

$$m = 10^5 = 100000 \; ; a = 321$$

$$c = 50001 \text{ ; and } u_0 = 38975$$

Then
$$\begin{aligned} u_1 &= (321\, u_0 + 50001) \bmod 100000 \\ &= (321 \times 38975 + 50001) \bmod 100000 \\ &= 12560976 \bmod 100000 = 60976 \\ u_2 &= (321 \times u_1 + 50001) \bmod 100000 \\ &= (321 \times 60976 + 50001) \bmod 100000 \\ &= 19623297 \bmod 100000 = 23297 \end{aligned}$$

Continuing the chain, we get

$$\begin{aligned} &u_3 = 28338, \quad u_4 = 46499, \quad u_5 = 76180, \\ &u_6 = 03781, \quad u_7 = 63702, \quad u_8 = 98343, \\ &u_9 = 18104, \quad u_{10} = 61385, \quad u_{11} = 54586, \end{aligned}$$

and $u_{12} = 72107$

Now, we can get uniform random numbers in the range 0 to 1 by assigning a decimal before every number. Thus, a sequence

$$\left\{ \begin{array}{l} 0.60976, 0.23297, 0.28338, 0.46499, 0.76180, 0.03781, \\ 0.63702, 0.98343, 0.18104, 0.61385, 0.54586, 0.72107 \end{array} \right\}$$

constitutes uniformly distributed random numbers in the interval 0 to 1.

Example 13.2. *Test the sequence of random numbers generated in the above example for randomness.*

Sol. Since the sample is very small we cannot conduct the frequency test. However, if the first and second moments about the origin are approximately equal to $\frac{1}{2}$ and $\frac{1}{3}$, respectively, then we can accept that the sample is from uniform distribution.

$$\text{First moment} = \text{Mean} = \bar{u} = \frac{1}{12} \sum_{i=1}^{12} u_i$$

$$= \frac{6.07298}{12} = 0.5060816 \simeq 0.50$$

$$\text{Second moment} = \frac{1}{12} \sum_{i=1}^{12} u_i^2 = \frac{3.9047925}{12}$$

$$= 0.3253993 \simeq 0.3333$$

Therefore we consider this sequence to be a good sample from uniform distribution.

Regarding the internal serial correlation, the sequence generated here is too small to construct a correlogram. However, the first and second serial correlation co-efficients obtained from Eq. (9.8) are found to be – 0.303 and – 0.256 respectively. The limits at 5%

tolerance level computed from Eqs. (9.9) and (9.10) for the first serial correlation coefficient are – 0.654 and 0.473 respectively. Similarly the tolerance limits for the second serial correlation coefficient are – 0.688 and 0.488 respectively.

Since the estimated first and second autocorrelation co-efficients are falling well with in their corresponding tolerance limits we can say that the generated sequence is free from internal correlation.

13.3. GENERATION OF NORMAL RANDOM NUMBERS

Generation of uniform random numbers alone is seldom required in hydrology as many variables such as annual run off and annual precipitation may follow normal distribution and variables measured on a seasonal basis are generally skewed. So, in the generation of hydrologic sequences we often require normal random numbers or random numbers in modified forms to have the desired skewness. Therefore, we shall now discuss the methods of generating normal random numbers. We have already seen that a random variable $X = \mu + \sigma Z$ follows a normal distribution with mean μ and standard deviation, σ, if Z is a standardised normal variable. Thus it is sufficient if we discuss the generation of standardised normal numbers with mean 0 and variance 1. Fortunately, the random numbers with $N(0, 1)$ can be generated very easily using any one of the following methods.

The Inverse Transformation Method. If Z is a standardised normal variable its cumulative distribution function $F(z)$ is given by

$$F(z) = \frac{1}{\sqrt{2\pi}} \int_{-\infty}^{z} e^{-t^2/2}\, dt.$$

where t has been used as a dummy variable to avoid confusion with the limit of integration. If we define a new random variable U such that $U = F(z)$, and if $h(u)$ is the probability density function of U, then we can obtain $h(u)$ using the principle of derived distribution explained in Sec. 2.9 from the relation.

$$h(u)\, du = f(z)\, dz$$

It may be noted that $U = F(z)$ is a one to one non-decreasing function. Also when $z = -\infty$, $U = F(-\infty) = 0$, and when $z = \infty$, $U = F(\infty) = 1$.

In other words while the range of z is $-\infty$ to ∞, the corresponding range of U is 0 to 1. Therefore, we have

$$h(u) = f(z)\frac{dz}{du} = \frac{f(z)}{(du/dz)}.$$

But since $U = F(z)$, we have that $du/dz = f(z)$. Hence

$$h(u) = \frac{f(z)}{f(z)} = 1.$$

This means that U is uniformly distributed over the range 0 to 1 with a constant ordinate equal to 1 as its probability density function.

With this result in view, a sequence of normally distributed random numbers may easily be generated by first generating a sequence of uniformly distributed random numbers and then obtaining the normal random numbers z_i corresponding to each u_i such that $u_i = F(z_i)$ or $z_i = F^{-1}(u_i)$, where F^{-1} is called the inverse function. This is the reason why this method is called *inverse transformation method.* The use of this method is somewhat constrained due to the fact that the function $F(z)$ is an intractable integral.

One approach would be to have the cumulative distribution function $F(z)$ tabulated at specified points and to arrive at the values of z situated in between these values by using a linear or polynomial interpolation. This procedure is known as *table look up* and hence, this method is also called *Table look-up method*. Because the extensive tabulation of $F(z)$ and the interpolation subroutines consume a lot of computer memory and time, other methods have been sought for the generation of normal random numbers, which are described below.

The central limit theorem method. Let $U_1, U_2, \ldots\ldots, U_m$ be m independent and uniformly distributed variables each in the range 0 to 1. Then each variable has a mean equal to 0.5 and variance equal to 1/12. Let us define a new random variable X such that $X = U_1 + U_2 + \ldots\ldots + U_m$. Then according to central limit theorem, for m large, X will follow normal distribution with mean given by $\mu_x = \mu_{u_1} + \mu_{u_2} + \ldots\ldots + \mu_{u_m}$ and variance given by $\sigma_x^2 = \sigma_{u_1}^2 + \sigma_{u_2}^2 + \ldots\ldots + \sigma_{u_m}^2$. Since $\mu_{u_1} = \mu_{u_2} = \ldots\ldots = \mu_{u_m} = 0.5$ and $\sigma_{u_1}^2 = \sigma_{u_2}^2 = \ldots\ldots = \sigma_{u_m}^2 = 1/12$, X will follow normal distribution with mean equal to 0.5 m and variance equal to $m/12$. If we define $Z = (X - m/2)/\sqrt{(m/12)}$, then Z will follow normal distribution with mean 0 and variance 1.

We can therefore obtain a chain of standardised normal random numbers z_j from a chain of uniformly distributed random numbers u_i from the following equation.

$$z_j = \frac{\sum_{i=1}^{m} (u_k - 0.5m)}{\sqrt{(m/12)}} \qquad j = 1, 2, \ldots\ldots \qquad \ldots(13.4)$$

where $k = (j-1)m + i$. Thus to generate n normal numbers by this method, one needs to generate $(n \times m)$ uniform random numbers.

The choice of m becomes a decision between the increased cost of generating large number of uniformly distributed random numbers and the corresponding nearness of z to normal distribution. For most purposes, the choice of $m = 6$ or $m = 12$ is made because of the corresponding computational simplicity.

For $m = 6$

$$z_j = \left[\sum_{i=1}^{6} u_k - 3 \right] . \sqrt{2} \qquad \text{...(13.5)}$$

where $k = 6(j-1) + i$

For $m = 12$

$$z_j = \sum_{i=1}^{12} (u_k - 6) \qquad \text{...(13.6)}$$

where $k = 12(j-1) + i$.

It has been observed that $m = 12$ is too small to yield a reasonable normal approximation. Another disadvantage of this method is that the tails of the normal distribution are not well represented.

Box-Muller method. The third and more exact method of generating normal random numbers known as the *Box-Muller method* is now described.

Let z_1 and z_2 be two independently and normally distributed random variables each with mean 0 and variance 1. Then the joint probability density function for Z_1 and Z_2 at any point (z_1, z_2) in the Eucledian plane is given by

$$\phi(z_1, z_2) = f(z_1)\, f(z_2)$$

$$= \frac{1}{\sqrt{2\pi}} e^{-z_1^2/2} . \frac{1}{\sqrt{2\pi}} e^{-z_2^2/2}$$

or

$$\phi(z_1, z_2) = \frac{1}{2\pi} e^{-(z_1^2 + z_2^2)/2} \qquad \text{...(13.7)}$$

Thus, the outcome of the underlying experiment produces a random point or a random vector (z_1, z_2) in the Eucledian space of two dimensions. Such a random point might be represented as well in polar coordinates (R, θ) constituting the transformation of the random variables Z_1 and Z_2 according to

$$z_1 = R \cos \theta \qquad \text{...(13.8)}$$

$$z_2 = R \sin \theta \qquad \text{...(13.9)}$$

The joint probability density function for the random variables R and 0, according to the principle of multivariate derived distributions explained in Section 2.9, then becomes

$$g(r, 0) = \phi\,(r \cos 0,\ r \sin 0) \left| \frac{\partial\,(z_1, z_2)}{\partial\,(r, 0)} \right|$$

where $$\left| \frac{\partial\,(z_1, z_2)}{\partial (r, 0)} \right| = \det \begin{bmatrix} \dfrac{\partial z_1}{\partial r} & \dfrac{\partial z_1}{\partial 0} \\ \dfrac{\partial z_2}{\partial r} & \dfrac{\partial z_2}{\partial \theta} \end{bmatrix} = \det \begin{bmatrix} \cos \theta & -r \sin \theta \\ \sin \theta & r \cos \theta \end{bmatrix}$$

$$= r \cos^2 \theta - (-r \sin^2 0) = r$$

$$g(r, 0) = \frac{r}{2\pi} e^{-(r^2 \cos^2 \theta + r^2 \sin^2 0)/2}$$

$$g(r, \theta) = \frac{r}{2\pi} e^{-r^2/2} \qquad \text{...(13.10)}$$

From this joint probability density function, the marginal density function for θ is evaluated as

$$f(\theta) = \int_{r=0}^{\infty} g(r, \theta)\, dr = \int_{r=0}^{\infty} \frac{r}{2\pi} e^{-r_2/2}\, dr$$

$$= \frac{1}{2\pi} \int_{r=0}^{\infty} e^{-r^2/2}\, d\left(\frac{r^2}{2}\right)$$

$$= \frac{1}{2\pi} \left[1 - e^{-r^2/2} \right]_0^{\infty} = \frac{1}{2\pi}$$

$$\left. \begin{aligned} f(\theta) &= \frac{1}{2\pi} \quad 0 \le \theta \le 2\pi \\ &= 0 \quad \text{otherwise} \end{aligned} \right\} \qquad \text{...(13.11)}$$

It means that θ has a uniform distribution in the range 0 to 2π. Similarly, the marginal density function for R becomes

$$h(r) = \int_0^{2\pi} g(r, \theta)\, d\theta = \int_0^{2\pi} \frac{r}{2\pi} e^{-r^2/2}\, d\theta = re^{-r^2/2}$$

$$\left. \begin{aligned} h(r) &= re^{-r^2/2} \quad r \ge 0 \\ &= 0 \quad \text{otherwise} \end{aligned} \right\} \qquad \text{...(13.12)}$$

The distribution for R given by Eq. (13.12) is known as Rayleigh distribution. Noteworthy is the result that the random variables R and θ are also independently distributed since

$$g(r, \theta) = f(\theta) \,.\, h(r) \quad \text{for all coordinates } (r, \theta).$$

Hence, if one can generate random variables θ and R having the respective distributions given by Eqs. (13.11) and (13.12), then the transformations $Z_1 = R \cos \theta$ and $Z_2 = R \sin \theta$ will yield independently distributed standardised normal variates Z_1 and Z_2.

The variable R with the probability density function given by Eq. (13.12) can be easily generated via the cumulative distribution transformation (or the inverse transformation). The cumulative distribution function of R is given by

$$H(r) = \int_0^r h(t)\, dt = \int_0^r te^{-t^2/2}\, dt$$

$$= \int_0^r e^{-t^2/2}\, d\left(\frac{t^2}{2}\right) = \left[1 - e^{-t^2/2}\right]_0^r$$

$$H(r) = 1 - e^{-r^2/2} \qquad \text{...(13.13)}$$

If U_1 is a uniformly distributed random variable in the range 0 to 1, then

$$U_1 = H(r) = 1 - e^{-r^2/2}$$

That is, $$R = [-2 \ln (1 - U_1)]^{1/2} \qquad \text{...(13.14)}$$

is the variable with required Rayleigh distribution. Similarly if U_2 is another uniformly distributed random variable in the range 0 to 1, then

$$\theta = 2\pi\, U_2 \qquad \text{...(13.15)}$$

will be distributed uniformly in the range 0 to 2π. Therefore, the variables defined by

$$\left.\begin{aligned} Z_1 &= R \cos \theta = [-2 \ln (1 - U_1)]^{1/2} \cos (2\pi U_2) \\ Z_2 &= R \sin \theta = [-2 \ln (1 - U_1)]^{1/2} \sin (2\pi\, U_2) \end{aligned}\right\} \qquad \text{...(13.16)}$$

will be two independently distributed standardised normal variables.

If U is uniformly distributed in the interval (0, 1), then $(1 - U)$ is also uniformly distributed in the same internal (0, 1). Hence some computational advantage may be gained by writing Eq. (13.16) in the form

$$\left.\begin{aligned} Z_1 &= (-2 \ln U_1)^{1/2} \cos (2\pi\, U_2) \\ Z_2 &= (-2 \ln U_1)^{1/2} \sin (2\pi\, U_2) \end{aligned}\right\} \qquad \text{...(13.17)}$$

Thus, we can generate a chain of standard normal random numbers from a chain of uniform random numbers taking the uniform random numbers in pairs each time from the following equations.

$$\left.\begin{aligned} z_i &= [-2 \ln (1 - u_i)]^{1/2} \cos (2\pi\, u_{i+1}) \\ z_{i+1} &= [-2 \ln (1 - u_i)]^{1/2} \sin (2\pi\, u_{i+1}) \end{aligned}\right\} \qquad \text{...(13.18)}$$

We prefer to use the form of Eq. (13.16) because it is possible that uniform random number u_i may take a value of zero in which case its logarithm becomes infinity and the computer gives overflow error when we use the form of Eq. (13.17). Please note that u_i will not take a value of 1. It may take a value of 0.99999 ......

Example 13.3. *Obtain the chain of normal random numbers using the table look-up method corresponding to the chain of uniform random number of Example 13.1.*

Sol. Corresponding to every random number u_i of the chain, a normal random number z_i is obtained such that $F(z_i) = u_i$, using linear interpolation from the CDF table of Normal distribution given in Table A-1 of the Appendix. For example, consider $u_1 = 0.60976$. From the table, we have that when

$$z = 0.26, F(z) = 0.6026$$

and when $$z = 0.27, F(z) = 0.6406$$

by linear interpolation when $z = 0.26188$, we get

$$F(z) = 0.60976$$

$$\therefore \quad z_1 = 0.26188.$$

Similarly, all other normal random numbers are obtained and tabulated below.

Table 13.1. Normal random numbers from Table look-up method

i	u_i	$z_i = F^{-1}(u_i)$	i	u_i	$z_i = F^{-1}(u_i)$
1	0.60976	0.26188	7	0.63702	0.35058
2	0.23297	– 0.71956	8	0.98343	2.15049
3	0.28338	– 0.57487	9	0.18104	– 0.91138
4	0.46499	– 0.08778	10	0.61385	0.28934
5	0.76180	0.71113	11	0.54586	0.11515
6	0.3781	– 1.77671	12	0.72107	0.58698

The reader may verify that the mean and standard deviation of these normal random numbers are 0.033 and 0.94, respectively, which are approximately equal to 0 and 1. The reader may also note that this sequence of normal random numbers has been used in Examples 11.1 and 11.4 while generating the synthetic streamflows.

Example 13.4. *Obtain two random normal numbers from the sequence of uniform random numbers of Example 13.1 using central limit theorem.*

Sol. $$z_1 = \left[\sum_{i=1}^{6} u_i - 3\right]\sqrt{2}$$

$$= (2.39071 - 3)\sqrt{2}$$
$$= -0.60929\sqrt{2} = -0.8617.$$

$$z_2 = \left[\sum_{i=7}^{12} u_i - 3\right]\sqrt{2}$$

$$= (3.68227 - 3)\sqrt{2}$$
$$= 0.68227\sqrt{2} = 0.9649.$$

Example 13.5. *Using the sequence of uniform random numbers of Example 13.1 in pairs, obtain the sequence of normal random numbers employing the Box-Muller method.*

Sol. The required computations are shown in Table 13.2.

Table 13.2. Normal random numbers using Box-Muller Method

u_1	u_2	$[-2\ln(1-u_1)]^{1/2}$	$2\pi u_2$	$\cos 2\pi u_2$	$\sin 2\pi u_2$	z_1	z_2
0.60976	0.23297	1.37185	1.46379	0.10680	0.99428	0.14651	1.36400
0.28338	0.46499	0.81634	2.92162	– 0.97590	0.21820	– 0.79667	0.17812
0.76180	0.37810	1.69390	2.37567	– 0.72074	0.69320	– 1.22086	1.17421
0.63702	0.98343	1.42366	6.17907	0.99458	– 0.10393	1.41594	– 0.14796
0.18104	0.61385	0.63201	3.85693	– 0.75487	– 0.65587	– 0.47708	– 0.41452
0.54586	0.72107	1.25646	4.53062	– 0.18077	– 0.98352	– 0.22713	– 1.23575

Thus, the sequence {0.14651, 1.364, – 0.79667, 0.17812, – 1.22086, 1.17421, 1.41549, – 0.14796, – 0.47708, – 0.41452, – 0.22713, – 1.23575}

constitute a chain of normal random numbers. The reader may verify that this sequence has a mean of 0.02 and a standard deviation of 0.9253.

13.4. RANDOM NUMBERS OF OTHER DISTRIBUTIONS

Sometimes random numbers following distributions other than normal may be required. The widely used method is the method of inverse transformation. This method can be used to transform uniformly distributed random numbers into random numbers of any distribution as long as the inverse, F^{-1}, of the desired distribution can be found analytically or approximately numerically.

The expressions given below may be used to obtain the random numbers following specified distributions. Only a few important cases are indicated. In these expressions u stands for uniformly distributed random number over the range 0 to 1 and z denotes the normal random number with zero mean and unit variance. Where more than one value are required, a typical value is designated as u_i or z_i. All values are taken independently of one another. y denotes the random number of desired distribution.

Exponential Distribution. If Y is following an exponential distribution with p.d.f.

$$f(y) = \lambda e^{-\lambda(y-\mu)}, \text{ with } \mu \le y \le \infty,$$

then the random numbers of this distribution can be generated from

$$y = -\frac{1}{\lambda} \ln (1-u) + \mu \qquad ...(13.19)$$

Gamma Distribution. If y is following a gamma distribution with p.d.f.

$$f(y) = \frac{\lambda\eta}{\Gamma\eta} e^{-\lambda y} y^{\eta - 1} \text{ with } 0 \le y \le \infty$$

and η an integer then, the random numbers following this distribution are generated from

$$y = -\frac{1}{\lambda} \sum_{i=1}^{\eta} \ln (1-u_i) \qquad ...(13.20)$$

Log-Normal Distribution. If Y is following log-normal distribution with p.d.f.

$$f(y) = \frac{1}{y\sigma\sqrt{2\pi}} e^{\frac{(\ln y - \mu)^2}{2\sigma^2}} \text{ with } 0 \le y \le \infty$$

Then the random numbers following this distribution are generated from

$$y = e^{\sigma z + \mu} \qquad ...(13.21)$$

Binomial Distribution. If Y is following binomial distribution with p.m.f.

$$p(y) = \binom{n}{y} p^y (1-p)^{n-y} \text{ ; with } y = 0, 1, 2, ..., n$$

then

$$y = \sum_{i=1}^{n} k_i \qquad ...(13.22)$$

will give the random numbers following this distribution, where

$$k_i = 0, \text{ if } u_i < p,$$

and

$$k_i = 1 \text{ if } u_i \ge p \text{ ;}$$

$$i = 1, 2, ..., n.$$

Poission Distribution. If Y is following Poisson distribution with p.m.f.

$$p(y) = \frac{e^{-\lambda} \lambda^y}{y\,!}, y = 0, 1, 2, \ldots\ldots$$

then the random numbers following this distribution can be obtained from

$$y = k \qquad \ldots(13.23)$$

where k is the lowest integer such that

$$\sum_{i=1}^{k+1} -\frac{1}{\lambda} \ln(1 - u_i) > 1$$

All these expressions are obtained through the appropriate cumulative distribution functions of the corresponding distributions.

EXERCISES

13.1. Explain the mixed congruential method of generating uniformly distribution random numbers in the range 0 to 1.

13.2. Describe the methods of generating standard normal random numbers by inverse transform and central limit theorem.

13.3. Describe the Box-Muller method of generating normal random numbers.

13.4. How do you select the parameters in mixed congruential algorithm for generating uniformly distributed random numbers to ensure minimum serial correlation and maximum periodicity.

13.5. Generate a chain of 12 uniformly distributed random numbers in the range 0 to 0.99999 using the mixed congruential method. The increment, multiplier, modulus and the seed may be taken as 54733, 321, 10^5 and 42875 respectively. Hence obtain two normal random numbers of $N(0, 1)$ using the central limit theorem method.

13.6. Obtain 12 standard normal random numbers using the 12 uniformly distributed random numbers obtained in Exercise 13.5 and table-look up method. Compute mean, standard deviation and first serial correlation coefficient for this chain.

13.7. Using the uniformly distributed random numbers of Exercise 13.5 and Box-Muller method, generate 12 standard normal random numbers. Compute mean, standard deviation and first serial correlation coefficient for this chain.

13.8. Generate a sequence of 500 normal random number on a computer. Determine the mean, standard deviation, correlogram upto 30 lags and the corresponding spectral density function. Verify whether the sequence represents a truly independent process.

Bibliography

1. Anderson, R.L. (1962) : *"Distribution of the Serial Correlation Co-efficient"*, Ann. Math Stat., Vol. 13, pp 1-63.

2. Bendat, J.S. and A.G. Piersol (1971) : *"Random Data Analysis and Measurement Procedures"*, Wiley-Interscience, New York.

3. Benjamin, J.R. and C.A. Cornell (1970) : *"Probability Statistics and Decision for Civil Engineers"*, Mc-Graw Hill, New York.

4. Box, G.E. and G.M. Jenkins, (1970) : *"Time Series, Analysis-Forecasting and Control"*, Holden Day, San Francisco.

5. Burges, S.J. (1977) : *"Some Problems with Log-Normal Markov Runoff Models"*, J. Hyd. Div., A.S.C.E., 98 (Hy 9), pp 1487-1496.

6. Chow, V.T., D.R. Maidment and L.W. Mays (1988) : *"Applied Hydrology"*, Mc-Graw Hill Book Company, New York.

7. Ciriani, T.A., Maione U, and J.R. Wallis, Editors, (1977) : *"Mathematical Models for Surface Water Hydrology"*, John Wiley and Sons, London.

8. Clarke, R.T. (1973) : *"Mathematical Models in Hydrology"*, Irrigation and Drainage Paper, 19, FAO, Rome.

9. Davenport, W.B. (1970) : *"Probability and Random Processes"*, Mc-Graw Hill New York.

10. Ecker, J.G. and M. Kupferschmid (1988) : *"Introduction to Operations Research"*, John Wiley and Sons, New York.

11. Fiering, M.B. (1967) : *"Streamflow Synthesis"*, Harvard University Press. Cambridge, Massachusettes.

12. Fiering, M.B. and B.B. Jackson (1971) : *"Synthetic Streamflows"*, Water Resources Monograph I, American Geophysical Union, Washington.

13. Freund, J.E. and R.E. Walpole (1987) *"Mathematical Statistics"*, Prentice Hall of India, New Delhi.

14. Gumbel, E.J. (1958) : *"Statistics of Extremes"*, Columbia University Press, New York.

15. Haan, C.T. : *"Statistical Methods in Hydrology"*, The Iowa State University Press, Ames.

16. Hahn, and Shapiro (1970) : *"Statistical Models in Engineering"*, John Wiley and Sons, New York.

17. Harvey, A.C. (1984) : *"Time Series Models"*, Heritage Publishers, New Delhi.

18. Hines, W.W. and D.C. Montgomery (1980) : *"Probability and Statistics in Engineering and Management Science. (Second edition)"* John Wiley and Sons, New York.

19. Jaya Rami Reddy, P (1978) : *"Optimal Operation of Reservoirs System in Godavari Basin"*, Ph. D. Thesis submitted to Kakatiya University.

20. Jenkins, G.M. and D.G. Watts (1968) : *"Spectral Analysis and its Applications"*, Holden-day, San Francisco.

21. Johnson, N.L. and S. Kotz (1972) : *"Distributions in Statistics"*, John Wiley and Sons, New York.

22. Kisiel, C.K. (1968) : *"Time Series Analysis"*, In Advances in Hydrosciences Vol 5, edited by V.T. Chow, Mc-Graw Hill, New York.

23. Kottegoda, N.T. (1980) : *"Stochastic Water Resources Technology"*, The Mac Millan Press Ltd., London.

24. Mass, A. et al (1962) : "*Design of Water Resource Systems*", Harvard University Press, Cambridge, Massachusettes.

25. Matalas, N.C. (1967) : "*Mathematical Assesment of Synthetic Hydrology,*" Water Resources Research, 3(4), pp 937-945.

26. Mihram, G (1972) : "*Simulation : Statistical Foundations and Methodology*", Academic Press, New York.

27. Quimpo, R.G. (1968) : "*Stochastic Analysis of Daily River Flows*", J. Hyd. Div., A.S.C.E., 94 (Hy 1), pp 43-57.

28. Robinson, E.A. (1980) : "*Physical Applications of Stationary Time Series*", Charless Griffin and Company Ltd. London.

29. Young G.K. and W.C. Pisano (1968) : "*Operational Hydrology Using Residuals*", J. Hyd. Div., A.S.C.E., 94 (Hy 4), pp 909-923.

30. Yevjewich, V. (1972) : "*Probability and Statistics in Hydrology*", Water Resources Publication, Fort Collins, Colorado,

31. Yevjewich, V. (1972) : "*Stochastic Processes in Hydrology*", Water Resouces Publication, Fort Collins, Colorado.

APPENDIX

(Statistical Tables)

TABLE A-1
Table of the Standard Cumulative Normal Distribution

$$F(z) = (2\pi)^{-1/2} \int_{-\infty}^{z} e^{-u^2/2} \, . du$$

z	.00	.01	.02	.03	.04	.05	.06	.07	.08	.09
.0	.5000	.5040	.5080	.5120	.5160	.5199	.5239	.5279	.5319	.5359
.1	.5398	.5438	.5478	.5517	.5557	.5596	.5636	.5675	.5714	.5753
.2	.5793	.5832	.5871	.5910	.5948	.5987	.6026	.6064	.6103	.6141
.3	.6179	.6217	.6255	.6293	.6331	.6368	.6406	.6443	.6480	.6517
.4	.6554	.6591	.6628	.6664	.6700	.6736	.6772	.6808	.6844	.6879
.5	.6915	.6950	.6985	.7019	.7054	.7088	.7123	.7157	.7190	.7224
.6	.7257	.7291	.7324	.7357	.7389	.7422	.7454	.7486	.7517	.7549
.7	.7580	.7611	.7642	.7673	.7703	.7734	.7764	.7794	.7823	.7852
.8	.7881	.7910	.7939	.7967	.7995	.8023	.8051	.8078	.8106	.8133
.9	.8159	.8186	.8212	.8238	.8264	.8289	.8315	.8340	.8365	.8389
1.0	.8413	.8438	.8461	.8485	.8508	.8531	.8554	.8577	.8599	.8621
1.1	.8643	.8665	.8686	.8708	.8729	.8749	.8770	.8790	.8810	.8830
1.2	.8849	.8869	.8888	.8907	.8925	.8944	.8962	.8980	.8997	.90147
1.3	.90320	.90490	.90658	.90824	.90988	.91149	.91309	.91466	.91621	.91774
1.4	.91924	.92073	.92220	.92364	.92507	.92647	.92785	.92922	.93056	.93189

z	.00	.01	.02	.03	.04	.05	.06	.07	.08	.09
1.5	.93319	.93448	.93574	.93699	.93822	.93943	.94062	.94179	.94295	.94408
1.6	.94520	.94630	.94738	.94845	.94950	.95053	.95154	.95254	.95352	.95449
1.7	.95543	.95637	.95728	.95818	.95907	.95994	.96080	.96164	.96246	.96327
1.8	.96407	.96485	.96562	.96638	.96712	.96784	.96856	.96926	.96995	.97062
1.9	.97128	.97193	.97257	.97320	.97381	.97441	.97500	.97558	.97615	.97670
2.0	.97725	.97778	.97831	.97882	.97932	.97982	.98030	.98077	.98124	.98169
2.1	.98214	.98257	.98300	.98341	.98382	.98422	.98461	.98500	.98537	.98574
2.2	.98610	.98645	.98679	.98713	.98745	.98778	.98809	.98840	.98870	.98899
2.3	.98928	.98956	.98983	$.9^{2}0097$	$.9^{2}0358$	$.9^{2}0613$	$.9^{2}0863$	$.9^{2}1106$	$.9^{2}1344$	$.9^{2}1576$
2.4	$.9^{2}1802$	$.9^{2}2024$	$.9^{2}2240$	$.9^{2}2451$	$.9^{2}2656$	$.9^{2}2857$	$.9^{2}3053$	$.9^{2}3244$	$.9^{2}3431$	$.9^{2}3613$
2.5	$.9^{2}3790$	$.9^{2}3963$	$.9^{2}4132$	$.9^{2}4297$	$.9^{2}4457$	$.9^{2}4614$	$.9^{2}4766$	$.9^{2}4915$	$.9^{2}5060$	$.9^{2}5201$
2.6	$.9^{2}5339$	$.9^{2}5473$	$.9^{2}5604$	$.9^{2}5731$	$.9^{2}5855$	$.9^{2}5975$	$.9^{2}6093$	$.9^{2}6207$	$.9^{2}6319$	$.9^{2}6427$
2.7	$.9^{2}6533$	$.9^{2}6636$	$.9^{2}6736$	$.9^{2}6833$	$.9^{2}6928$	$.9^{2}7020$	$.9^{2}7110$	$.9^{2}7197$	$.9^{2}7282$	$.9^{2}7365$
2.8	$.9^{2}7445$	$.9^{2}7523$	$.9^{2}7599$	$.9^{2}7673$	$.9^{2}7744$	$.9^{2}7814$	$.9^{2}7882$	$.9^{2}7948$	$.9^{2}8012$	$.9^{2}8074$
2.9	$.9^{2}8134$	$.9^{2}8193$	$.9^{2}8250$	$.9^{2}8305$	$.9^{2}8359$	$.9^{2}8411$	$.9^{2}8462$	$.9^{2}8511$	$.9^{2}8559$	$.9^{2}8605$
3.0	$.9^{2}8650$	$.9^{2}8694$	$.9^{2}8736$	$.9^{2}8777$	$.9^{2}8817$	$.9^{2}8856$	$.9^{2}8893$	$.9^{2}8930$	$.9^{2}8965$	$.9^{2}8999$
3.1	$.9^{3}0324$	$.9^{3}0646$	$.9^{3}0957$	$.9^{3}1260$	$.9^{3}1553$	$.9^{3}1836$	$.9^{3}2112$	$.9^{3}2378$	$.9^{3}2636$	$.9^{3}2886$
3.2	$.9^{3}3129$	$.9^{3}3363$	$.9^{3}3590$	$.9^{3}3810$	$.9^{3}4024$	$.9^{3}4230$	$.9^{3}4429$	$.9^{3}4623$	$.9^{3}4810$	$.9^{3}4991$
3.3	$.9^{3}5166$	$.9^{3}5335$	$.9^{3}5499$	$.9^{3}5658$	$.9^{3}5811$	$.9^{3}5959$	$.9^{3}6103$	$.9^{3}6242$	$.9^{3}6376$	$.9^{3}6505$
3.4	$.9^{3}6631$	$.9^{3}6752$	$.9^{3}6869$	$.9^{3}6982$	$.9^{3}7091$	$.9^{3}7197$	$.9^{3}7299$	$.9^{3}7398$	$.9^{3}7493$	$.9^{3}7585$
3.5	$.9^{3}7674$	$.9^{3}7759$	$.9^{3}7842$	$.9^{3}7922$	$.9^{3}7999$	$.9^{3}8074$	$.9^{3}8146$	$.9^{3}8215$	$.9^{3}8282$	$.9^{3}8347$
3.6	$.9^{3}8409$	$.9^{3}8469$	$.9^{3}8527$	$.9^{3}8583$	$.9^{3}8637$	$.9^{3}8689$	$.9^{3}8739$	$.9^{3}8787$	$.9^{3}8834$	$.9^{3}8879$
3.7	$.9^{3}8922$	$.9^{3}8964$	$.9^{4}0039$	$.9^{4}0426$	$.9^{4}0799$	$.9^{4}1158$	$.9^{4}1504$	$.9^{4}1838$	$.9^{4}2159$	$.9^{4}2468$
3.8	$.9^{4}2765$	$.9^{4}3052$	$.9^{4}3327$	$.9^{4}3593$	$.9^{4}3848$	$.9^{4}4094$	$.9^{4}4331$	$.9^{4}4558$	$.9^{4}4777$	$.9^{4}4988$
3.9	$.9^{4}5190$	$.9^{4}5385$	$.9^{4}5573$	$.9^{4}5753$	$.9^{4}5926$	$.9^{4}6092$	$.9^{4}6253$	$.9^{4}6406$	$.9^{4}6554$	$.9^{4}6696$

TABLE A-2

Percentage Points of Chi-Square Distribution

Value of $\chi^2_{\upsilon,\alpha}$ such that Prob. $[\chi_\upsilon^{\,2} > \chi^2_{\upsilon,\alpha}] = \alpha$

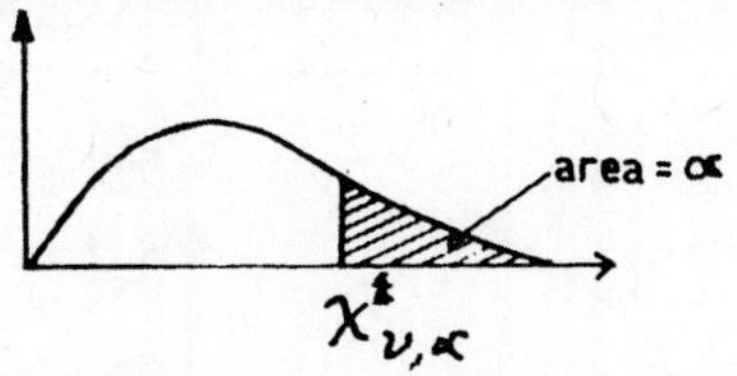

υ	0.995	0.990	0.975	.0950	0.900	0.10	0.05	0.025	0.010	0.005
1	0.000039	0.00016	0.00098	0.0049	0.0158	2.71	3.84	5.02	6.63	7.788
2	0.0100	0.0201	0.0506	0.103	0.211	4.16	5.99	7.38	9.21	10.60
3	0.0717	0.115	0.216	0.253	0.584	6.25	7.81	9.35	11.34	12.84
4	0.207	0.297	0.484	0.711	1.06	7.78	9.49	11.14	13.28	14.86
5	0.412	0.554	0.831	1.15	1.61	9.24	11.07	12.83	15.09	16.75
6	0.676	0.872	1.24	1.64	2.20	10.64	12.59	14.45	16.81	18.55
7	0.989	1.24	1.69	2.17	2.83	12.02	14.07	16.01	18.48	20.28
8	1.34	1.65	2.18	2.73	3.49	13.36	15.51	17.53	20.09	21.96
9	1.73	2.09	2.70	3.33	4.17	14.68	16.92	19.02	21.67	23.59
10	2.16	2.56	3.25	3.94	4.87	15.99	18.31	20.48	23.21	25.19
11	2.60	3.05	3.82	4.57	5.58	17.28	19.68	21.92	24.73	26.76
12	3.07	3.57	4.40	5.23	6.30	18.55	21.03	23.34	26.22	28.30

υ	*0.995*	*0.990*	*0.975*	*0.950*	*0.900*	*0.10*	*0.05*	*0.025*	*0.010*	*0.005*
13	3.57	4.11	5.01	5.89	7.04	19.81	22.36	24.74	27.69	29.82
14	4.07	4.66	5.63	6.57	7.79	28.06	23.68	26.12	29.14	31.32
15	4.60	5.23	6.26	7.26	8.55	22.31	25.00	27.49	30.58	32.80
16	5.14	5.81	6.91	7.96	9.31	23.54	26.30	28.85	32.00	34.27
17	5.70	6.41	7.56	8.67	10.08	24.77	27.59	30.19	33.41	35.72
18	6.26	7.01	8.23	9.39	10.86	25.99	28.87	31.53	34.81	37.16
19	6.84	7.63	8.91	10.12	11.65	27.20	30.14	32.85	36.19	38.58
20	7.43	8.26	9.59	10.85	12.44	28.41	31.41	34.17	37.57	40.00
21	8.03	8.90	10.28	11.59	13.24	29.62	32.67	35.48	38.93	41.40
22	8.64	9.54	10.98	12.34	14.04	30.81	33.92	36.78	40.29	42.80
23	9.26	10.20	11.69	13.09	14.85	32.01	35.17	38.08	41.64	44.18
24	9.89	10.86	12.40	13.85	15.66	33.20	36.42	39.36	42.98	45.56
25	10.52	11.52	13.12	14.61	16.47	34.38	37.65	40.65	44.31	46.93
26	11.16	12.20	13.80	15.38	17.29	35.56	38.88	41.92	45.64	48.29
27	11.81	12.88	14.57	16.15	18.11	36.74	40.11	43.19	46.96	49.64
28	12.46	13.56	15.31	16.93	18.94	37.92	41.34	44.46	48.28	50.99
29	13.12	14.26	16.05	17.71	19.77	39.09	42.56	45.72	49.59	52.34
30	13.79	14.95	16.79	18.49	20.60	40.26	43.77	46.98	50.89	53.67

For larger degrees of freedom ($\upsilon > 30$) χ_{υ}^{2} is approximately normal with mean = υ and variance = 2υ.

TABLE A-3

Critical Value Δ_0 of the Smirnov-Kolmogorov Statistic Δ for Various Values of 'n', and 'α' Currently used in Hydrology

n	α = *0.20*	*0.10*	*0.05*	*0.01*
5	0.45	0.51	0.56	0.67
10	0.32	0.37	0.41	0.49
15	0.27	0.30	0.34	0.40
20	0.23	0.26	0.29	0.36
25	0.21	0.24	0.27	0.32
30	0.19	0.22	0.24	0.29
35	0.18	0.20	0.23	0.27
40	0.17	0.19	0.21	0.25
45	0.16	0.18	0.20	0.24
50	0.15	0.17	0.19	0.23
$n > 50$	$\frac{1.07}{\sqrt{n}}$	$\frac{1.22}{\sqrt{n}}$	$\frac{1.36}{\sqrt{n}}$	$\frac{1.63}{\sqrt{n}}$

TABLE A-4

Mean and Standard Deviation of Reduced Extremes for Gumbel's Distribution

n	$\bar{y}_n$	σ_n	n	$\bar{y}_n$	σ_n
8	.4843	.9043	49	.5481	1.1590
9	.4902.	.9288	50	.54854	1.16066
10	.4952	.9497	51	.5489	1.1623
11	.4996	.9676	52	.5493	1.1638
12	.5035	.9833	53	.5497	1.1653
13	.5070	.9972	54	.5501	1.1667
14	.5100	1.0095	55	.5504	1.1681
15	.5128	1.02057	56	.5508	1.1696
16	.5157	1.0316	57	.5511	1.1708
17	.5181	1.0411	58	.5515	1.1721
18	.5202	1.0493	59	.5518	1.1734
19	.5220	1.0566	60	.55208	1.17467
20	.52355	1.06283	62	.5527	1.1770
21	.5252	1.0696	64	.5533	1.1793
22	.5268	1.0754	66	.5538	1.1814
23	.5283	1.0811	68	.5543	1.1834
24	.5296	1.0864	70	.55477	1.18536
25	.53086	1.09145	72	.5552	1.1873

n	$\bar{y}_n$	σ_n	n	$\bar{y}_n$	σ_n
26	.5320	1.0961	74	.5557	1.1890
27	.5332	1.1004	76	.5561	1.1906
28	.5343	1.1047	78	.5565	1.1923
29	.5353	1.1086	80	.55688	1.19382
30	.53622	1.11238	82	.5572	1.1953
31	.5371	1.1159	84	.5576	1.1967
32	.5380	1.1193	86	.5580	1.1980
33	.5388	1.1226	88	.5583	1.1994
34	.5396	1.1255	90	.5586	1.20073
35	.54034	1.12847	92	.5589	1.2020
36	.5410	1.1313	94	.5592	1.2032
37	.5418	1.1339	96	.5595	1.2044
38	.5424	1.1363	98	.5598	1.2055
39	.5430	1.1388	100	.56002	1.20649
40	.54362	1.14132	150	.56461	1.22534
41	.5442	1.1436	200	.56715	1.23598
42	.5448	1.1458	250	.56878	1.24292
43	.5453	1.1480	300	.56993	1.24786
44	.5458	1.1499	400	.57144	1.25450
45	.5463	1.15185	500	.57240	1.25880
46	.5468	1.1538	750	.57377	1.26506
47	.5473	1.1557	1000	.57450	1.26851
48	.5477	1.1574	∞	.57722	1.28255

Table A-5

Frequency Factor K for Use with Log Pearson Type III Distribution

Co-efficient of skew g	*Recurrence interval in years*							
	2	*5*	*10*	*25*	*50*	*100*	*200*	*1000*
	Per cent chance of exceedence							
	50	*20*	*10*	*4*	*2*	*1*	*0.5*	*0.1*
3.0	– 0.396	0.420	1.180	2.278	3.152	4.501	4.970	7.250
2.5	– 0.360	0.518	1.250	2.262	3.048	3.845	4.652	6.600
2.2	– 0.330	0.574	1.284	2.240	2.970	3.705	4.444	6.200
2.0	– 0.307	0.609	1.302	2.219	2.912	3.605	4.298	5.910
1.8	– 0.282	0.643	1.318	2.193	2.848	3.499	4.147	5.660
1.6	– 0.254	0.675	1.329	2.163	2.780	3.388	3.990	5.390
1.4	– 0.225	0.705	1.337	2.128	2.706	3.271	3.828	5.110
1.2	– 0.195	0.732	1.340	2.087	2.626	3.149	3.661	4.820
1.0	– 0.164	0.758	1.340	2.043	2.542	3.022	3.489	4.540
0.9	– 0.148	0.769	1.339	2.018	2.498	2.957	3.401	4.395
0.8	– 0.132	0.780	1.336	1.998	2.453	2.891	3.312	4.250
0.7	– 0.116	0.790	1.333	1.967	2.407	2.824	3.223	4.105
0.6	– 0.099	0.800	1.328	1.939	2.359	2.755	3.132	3.960

Co-efficient of skew g	Recurrence interval in years							
	2	5	10	25	50	100	200	1000
	Per cent chance of exceedence							
	50	20	10	4	2	1	0.5	0.1
0.5	– 0.083	0.808	1.323	1.910	2.311	2.686	3.041	3.815
0.4	– 0.066	0.816	1.317	1.880	2.261	2.615	2.949	3.670
0.3	– 0.050	0.824	1.309	1.849	2.211	2.544	2.856	3.525
0.2	– 0.033	0.830	1.301	1.818	2.159	2.472	2.763	3.380
0.1	– 0.017	0.836	1.292	1.785	2.107	2.400	2.670	3.230
0	0	0.842	1.282	1.751	2.054	2.326	2.576	3.090
– 0.1	0.017	0.836	1.270	1.716	2.000	2.252	2.482	2.950
– 0.2	0.033	0.850	1.258	1.680	1.945	2.178	2.288	2.810
– 0.3	0.050	0.853	1.245	1.643	1.890	2.104	2.394	2.675
– 0.4	0.066	0.855	1.231	1.606	1.834	2.029	2.201	2.540
– 0.5	0.083	0.856	1.216	1.567	1.777	1.955	2.108	2.400
– 0.6	0.099	0.857	1.200	1.528	1.720	1.880	2.016	2.275
– 0.7	0.116	0.857	1.183	1.488	1.663	1.806	1.926	2.150
– 0.8	0.132	0.856	1.166	1.448	1.606	1.733	1.837	2.035
– 0.9	0.148	0.854	1.147	1.407	1.549	1.660	1.749	1.910
– 1.0	0.164	0.852	1.128	1.366	1.492	1.588	1.664	1.800
– 1.2	0.195	0.844	1.086	1.282	1.379	1.449	1.501	1.625
– 1.4	0.225	0.832	1.041	1.198	1.270	1.318	1.351	1.465
– 1.6	0.254	0.817	0.994	1.116	1.166	1.197	1.216	1.280
– 1.8	0.282	0.799	0.945	1.035	1.069	1.087	1.097	1.130
– 2.0	0.307	0.777	0.895	0.959	1.980	0.990	0.995	1.000
– 2.2	0.330	0.752	0.844	0.888	0.900	0.905	0.907	0.910
– 2.5	0.360	0.711	0.771	0.793	0.798	0.799	0.800	0.802
– 3.0	0.396	0.636	0.660	0.666	0.666	0.667	0.667	0.668

Index

A

Autocorrelation coefficient, 139
Axioms of probability, 14

B

Bayes' theorem, 22
Bernoulli trials, 74
Beta distribution, 67
Binomial distribution, 74
Bivariate distribution, 30

C

Catastrophic event, 147
Central limit theorem, 59
Chebyshev's inequality, 44
Chi-square distribution, 66
Chi-square test, 103
Coefficient of determination, 118
Coefficient of multiple correlation, 127
Coefficient of variation, 44
Compliment set, 11
Conditional distribution, 33
Conditional probability, 18
Continuous spectrum, 176
Correlation coefficient, 47
Correlogram, 152
Covariance, 46
Covariance stationarity, 137
Cross covariance matrix, 221
Cumulative distribution function, 24

D

Definition of independence, 19
Derived distribution, 35
Descrete spectrum, 173
Dirac-delta function, 188
Disjoint set, 12
Distribution function, 24

E

Ensemble, 138
Ergodic process, 138
Erlang distribution, 64
Event, 7
Exponential distribution, 66
Extreme value distribution, 68

F

Fractional Gaussian noise (fGn) model, 203
Frequency factors, 108

G

Gamma distribution, 64
Gaussian distribution, 55
Geometric distribution, 80
Goodness of fit, 91
Gumbel distribution, 68

H

Harmonic, 172
Hurst phenomenon, 203
Hypergeometric distribution, 83

I

Identity set, 8
Independent process, 152
Interpretations of probability, 15
Intersection, 11

J

Joint probability function, 30
Jump, 145

K

Kurtosis coefficient, 50

L

Linear trend, 145
Line spectrum, 172
Log-normal distribution, 61

M

Marginal distribution, 32
Memory of the process, 192
Method of maximum likelihood, 89
Method of moments, 89
Mixed random variable, 28
Moment generating function, 52
Multiple regression, 127
Multivariate distribution, 30
Mutually exclusive set, 12

N

Negative binomial distribution, 79
Normal distribution, 55
Null set, 10
Nyquist frequency, 178

O

Order, 193

P

Pascal distribution, 79
Pearson distributions, 71
Periodogram, 173
Persistence, 192
Plotting position formulae, 94
Poisson distribution, 81
Polynomial fitting, 126
Probability density function, 24
Probability mass function, 24
Pseudorandom numbers, 232

R

Random number generator, 232
Random variable, 23
Random seed, 232
Range, 203
Rank, 93
Rank correlation coefficient, 120
Reduced variate, 69
Regression, 122
Relative compliment, 12

S

Sample points, 6
Sample space, 7
Scatter diagram, 116
Seed, 232
Sequent peak algorithm, 3
Set difference, 12
Shot noise model, 215
Skewness coefficient, 49
Smirnov-Kolmogorov test, 107
Smoothing windows, 178
Spectral density function, 177
Spurious correlation, 118
Stationarity in wide sense, 138
Stationary process, 137
Strictly stationary process, 137
Strong stationarity, 137

T

Table look-up method, 238
Test for stationarity, 140
Total probability theorem, 22

U

Uniform distribution, 67
Union, 10

V

Vacant set, 10
Variance density spectrum, 177
Variance spectrum, 173

W

Warming-up, 195
Weakly stationary process, 137
Weibull distribution, 70
White noise process, 180

Y

Yule-walker equations, 159